ENERGY SCIENCE, ENGINEERING AND TECHNOLOGY

NEW RESEARCH IN MICROWAVE PROCESSING OF CONCRETE

ENERGY SCIENCE, ENGINEERING AND TECHNOLOGY

Additional books in this series can be found on Nova's website under the Series tab.

Additional e-books in this series can be found on Nova's website under the eBooks tab.

ENERGY SCIENCE, ENGINEERING AND TECHNOLOGY

NEW RESEARCH IN MICROWAVE PROCESSING OF CONCRETE

NATT MAKUL

nova
science publishers
New York

Library of Congress Cataloging-in-Publication Data

ISBN: 978-1-53612-023-3

Published by Nova Science Publishers, Inc. † New York

To my Mom, Thongkum Makul

CONTENTS

PREFACE

Microwave heating is a highly efficient technique for various thermal processes. Advantages of microwave heating compared to conventional processing methods include energy-saving rapid heating rates and short processing times, deep penetration of the microwave energy, instantaneous and precise electronic control, clean heating processes, and no generation of secondary waste. Microwave energy processes for heating have been developed for numerous laboratory-scale investigations and, in some cases, have been commercialized. Microwave energy use should theoretically be advantageous in the processing of cement and concrete materials. These materials exhibit excellent dielectric properties and, therefore, should be able to absorb microwave energy very efficiently and instantaneously convert it into heat.

Theoretically, MW energy can be applied to cementitious (CM)-based materials. Water, which is a main component that reacts with cement particles in a hydration reaction, has a dielectric permittivity at room temperature that is higher than the dielectric permittivity of concrete's other constituents. Consequently, when an electric field is penetrated by the concrete's components, energy is transferred from the electromagnetic (electric) field to the water's molecular bonds (which comprise MW-irradiated materials) via friction among the positive and negative ions in compressed and uncompressed conditions. As a result, any heat that is generated is dissipated and transferred within the concrete to be processed, and the temperature of the concrete is elevated. Thus, an accelerated hydration reaction occurs. Instantaneously, free water molecules attach to the inside of the internal concrete structure before the cement hardens, which can collapse the capillary pores and generate a densified internal structure. Thus, the compressive strength development of the high-performance concrete (HPC) significantly increases, which is a main goal of the development of HPC for contemporary use.

This book presents a comprehensive study on the interaction between steel-reinforced concrete and microwave energy using a single-mode rectangular waveguide. We begin by investigating the mechanisms of changes in the adaptive dielectric properties of concretes and reinforced concretes to predict how these properties are altered when microwave energy is applied. Next, the structural characteristics of microwave-cured concrete will be identified. Finally, we will formulate mathematical models to describe the relationship between microwave curing and mass transfer for steel-reinforced concrete. The obtained results show that dielectric properties are relatively high and remain constant during the dormant period. After this period, the hydration reaction resumes and dielectric properties decrease rapidly. Further with the use of microwave heating, early-age strength increases during the first 7 days; however, during the next 7 days, early-age strength decreases slightly, until it reaches its lowest at the 28-day mark. The temperature rise as actually recorded at the centre of the sample during microwave heating in our experiment consistently agreed with figures calculated by a mathematical model.

This book consists of five chapters: Chapter One: "Advances in Microwave Heating"; Chapter Two: "Recent research and development"; Chapter Three: "Dielectric Properties of Concrete"; Chapter Four: "Microwave Heating of Concrete and Reinforced

Concrete"; Chapter Five: "A Theoritical Model to Predict The Properties of The Reinforced Concrete when Subjected to Microwave Energy"; and Chapter Six: "Conclusions and recommendations for Future Study" Each chapter comprises an introduction, a discussion of the concept of the design and the concrete's development, and the properties and testing of the concrete in fresh and hardened stages.

I appreciate the financial support of the Thailand Research Fund (Contract No. TRG5780255). The author is also grateful for the permission and facilities extended by Professor Dr. Phadungsak Rattanadecho at the Center of Excellence in Electromagnetic Energy Utilization in Engineering (CEEE) at Thammasat University in Thailand.

Natt Makul
Assistant Professor
Department of Building Technology
Faculty of Industrial Technology
Phranakhon Rajabhat University
Email: shinomomo7@gmail.com

Chapter 1

ADVANCES IN MICROWAVE HEATING

1.1. RESEARCH PROBLEM AND ITS SIGNIFICANCE

Curing by microwave irradiation (or microwave curing) is a high-performance curing method for dielectric materials used in a variety of industrial processes. On average, it is 1.50-2.0 times more effective than conventional curing methods. The primary differences between conventional curing methods and microwave curing are the heat generation mechanism and whether the direction of heat transfer is inward or outward. In the former method, the component being cured is heated by an external source and heat transfer is from the outside inward (Figure 1.1(a)). Microwave curing (Figure 1.1(b)) is based on the rapid polarization and depolarization of charged groups when the material to be heated is subjected to a high-frequency electromagnetic field, resulting in internal heat generation.

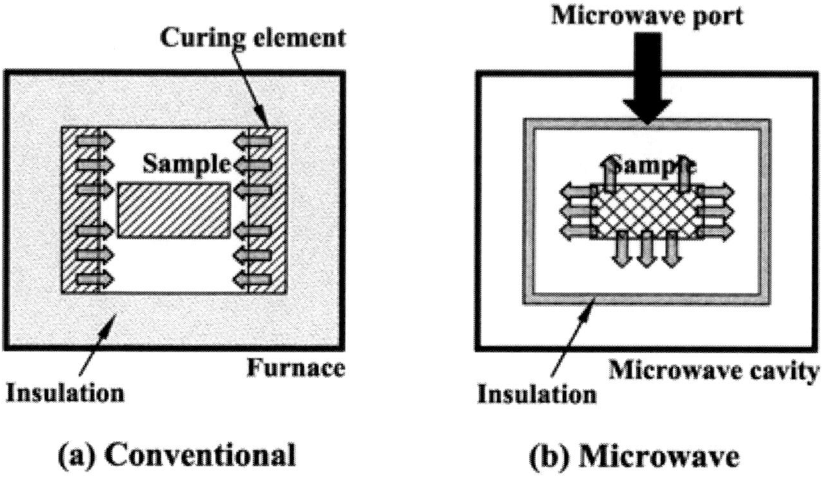

Figure 1.1. Comparison of heat generation and transfer mechanism in (a) conventional, and (b) microwave curing.

Microwave curing has been used in research and industrial applications such as thawing frozen meat, vulcanizing rubber, curing adhesive agents in lumber manufacture, curing of food, drying, decontamination of surfaces, and disinfection and sterilization. In order to be suitable for microwave curing the material in question must be a dielectric, and most current applications involve porous media composed of a solid phase containing a second fluid phase within the internal voids. Concrete may be classed as one of these materials.

Concrete is the predominant modern construction material with annually ½ cubic meters used per capita worldwide. One disadvantage to concrete is the slow development of strength under normal temperatures and pressures. Various accelerated-curing methods have been invented to enhance the rate of strength development, including application of external heat and pressure in an autoclave. However, the low thermal conductivity (0.8 to 1.28 W/m.K) of concrete reduces heat transfer during curing. Microstructural development in concrete is dependent on the uniformity of heat liberation from hydration and from external heating, and a steep temperature gradient from the external heat source to the inner regions of the component results in crack formation within the internal structure. In particular, concrete structures containing reinforcing steel are susceptible to fracture, substantially reducing the service life of the component. Thermal gradient problems may be reduced by using an internal heating process such as microwave curing.

Microwave curing of plain and steel-reinforced concrete is a relatively new field that will expand in importance in the near future. In particular, research is necessary to understand curing mechanisms with adaptive dielectric properties and to characterize the heat and mass transfer properties of concrete materials.

1.2. OBJECTIVES

This research aims to:

- Investigate the mechanisms of changes in the dielectric properties of plain containing pozzolan materials and reinforced concrete.
- Characterize the microstructure of microwave-cured plain and reinforced concrete.
- Study the mechanism of microwave heat generation, heat dissipation, and mass transfer of plain containing pozzolan materials and reinforced concrete.
- Formulate mathematical models for the relationships between microwave curing and mass transfer for steel-reinforced concrete.

1.3. Scope of Research

This research combines both experimental and theoretical analyses. The dielectric properties of plain and steel-reinforced concrete including the relative dielectric constant ε' and relative dielectric loss ε'' during the first 24 h of hydration were measured at a frequency of 2.45 ± 0.05 GHz, and the characteristics of concrete subjected to short-term microwave curing (without loss of moisture) were investigated experimentally and numerically.

To measure the dielectric properties of concrete, a network analyzer with an open-ended coaxial probe was used. The apparatus consisted of a coaxial cavity, a microwave reflectometer, a 3.5-mm coaxial cable, a 3.5-mm female calibration, and short-, open-, and matched-load software. The coaxial cavity was suitable for measurements from 1.5 - 2.6 GHz with a precision of better than 2% for the dielectric constant and 5% for the dielectric loss factor. In order to eliminate the effects of the metal thermocouple on the microwave distribution the temperature and dielectric properties were measured in three separate samples each.

The microwave system was a monochromatic magnetron source with a frequency of 2.45 GHz. The energy was transmitted directly along the propagation direction (+z) of a rectangular wave guide toward a water load situated at the end of the waveguide to ensure that a minimal amount of microwave energy would be reflected back to the sample. The heated water load was recirculated through a cooling tower.

The waveguide and auxiliary equipment limited direct access to the sample, and evaporation within the structure of the concrete could not be continuously measured. Instead, evaporation measurements were based on weight loss of the sample.

The samples were fabricated from Type I Portland cement and were 109.22 mm long, 54.61 mm wide, and 50 mm or 80 mm thick. Water-cement ratios by mass were 0.25, 0.38 and 0.45. After mixing, the concrete samples were moulded and covered with plastic to prevent moisture loss. The hardened concrete samples were demoulded after ½ h for microwave curing and 23½ h for water curing. In reinforced concrete samples, the reinforcing steel was arranged at ½ times of the thickness of the concrete sample.

The effects of controlled curing methods such as lime-saturation and air at 25 ± 2°C were experimentally investigated by monitoring the quantity and quality of hydration reaction products such as calcium silicate hydrate (C-S-H), and measuring changes in porosity and free and fixed moisture content in the cured samples.

During microwave curing, each sample was subjected to microwave energy in a rectangular waveguide for a specific amount of time. The microwave-cured samples were tested to determine compressive strength, curing water and air content, quantity and quality of hydration products such as calcium silicate hydrate (C-S-H), porosity, and changes in free and fixed moisture content.

Solutions to Maxwell's equations and the heat and mass transfer equations were used to understand heat and mass transfer phenomena occurring during microwave curing. The concrete was assumed to be a porous material with variable internal fluid flow. To simplify the problem, a 2-dimensional theoretical model was constructed to analyze the microwave behaviour using a TE_{10} propagation mode. Rattanadecho previously described a 3D model of the electromagnetic field inside a rectangular waveguide in TE_{10} mode developed using the COMSOL Multiphysics modeling software.

1.4. Expected Benefits

The benefits to be gained from this project are as follows:

- **Benefit to academic**

  The results of this research will lead to an in-depth understanding of the adaptive dielectric properties of plain and steel-reinforced concrete, enabling prediction of changes occurring during application of microwave energy and during hydration. Suitable mathematical models of heat and mass transfer and microstructural development during hydration and microwave curing will be developed. Results obtained using the laboratory scale apparatus may be applied to industrial curing systems for concrete, related cementitious materials, and eventually to microwave curing of other reactively-cured materials.

- **Benefit to faculty and related institution**

  This project will benefit higher education as well as providing technology transfer to cement and concrete production. The results will be published in high-impact peer-reviewed journals at an international level. Consequently, publication of this research may raise the standard of research at Thailand and Phranakorn Rajabhat University to an international level.

Chapter 2

RECENT RESEARCHES AND DEVELOPMENTS

2.1. INTRODUCTION

Cement and concrete products are well accepted as man-made construction materials. Owing to their numerous advantages (e.g., easy handling of compression forces, brittle nature that enables a rigid structure to be obtained, transportability, and durability/ability to withstand severe environments), these materials are used widely in civilian buildings and universally in infrastructures. A report on global cement use published by Morgan Stanley research (Table 2.1) found that worldwide cement consumption decreased in the 2006 –2013 period due to the economic downturn. The years 2012 and 2014 showed the slowest growth for every country in the world, except China and India, since the crisis years of 2007 –2009. However, cement consumption is expected to increase gradually from 2014 forwards.

The increased production of cement and concrete leads to the increased consumption of energy. Conventionally, the cement production procedure includes a clinkering process, consisting of precalcination and pyro-processing steps, as shown in Figure 2.1. This process requires energy consumption, especially in the cyclone preheater, precalciner (cyclone preheater plus calciner), and rotary kiln steps (burning consumes 60 – 70% of the fuel input). The average specific thermal energy consumption in a conventional cement clinker manufacturing process is about 31.5 GJ per ton of clinker.

Table 2.1. Cement consumption growth by region (year over year) (%)

Region	*Year*									
	2006	*2007*	*2008*	*2009*	*2010*	*2011*	*2012*	*2013E*[*]	*2014E*	*2015E*
Western Europe	5.0	0.0	-9.8	-19.5	-6.3	-0.5	-14.2	-6.8	-1.1	1.7
Eastern Europe	15.7	7.6	0.2	-12.6	3.0	8.6	-4.4	2.9	1.5	3.5
Former Soviet Union	13.8	16.8	0.2	23.4	9.5	13.9	9.2	4.3	3.8	4.1
North America	-0.3	-7.9	-15.8	-26.3	0.9	2.5	8.8	3.8	6.4	6.2
Latin America	11.8	8.5	6.4	-2.0	5.9	7.0	3.9	2.1	3.7	3.5
Middle East and North Africa (MENA)	7.6	11.8	10.2	7.8	6.6	1.6	4.1	1.4	3.9	5.1
Sub-Saharan Africa	11.8	15.7	1.9	5.3	3.3	11.4	9.6	7.0	6.8	7.1
China	14.9	13.1	3.3	18.3	14.4	10.8	6.6	7.1	3.5	2.1
India	12.2	8.9	5.0	10.6	7.2	6.2	6.7	3.6	5.2	6.8
North Asia	2.5	-0.8	-4.3	-8.9	2.5	2.0	1.8	3.3	0.0	0.6
South Asia	4.3	5.7	7.5	6.0	8.3	6.3	7.1	2.4	3.7	5.1
Australia/Pacific	-0.5	6.5	3.8	-8.5	1.0	4.9	-2.1	-0.8	3.0	-4.1
World (excl-China)	7.3	5.5	0.4	-4.7	4.0	5.1	3.1	2.1	3.5	4.6
Emerging Markets (excl-China)	9.8	9.3	5.1	0.7	6.2	6.1	4.9	2.8	3.9	4.9

Remark: Source from industry and national sources, Global Cement Report, Morgan Stanley Research estimates (E).

High-level energy consumption, which refers to both a high level of consumption and a long time for the burning process, is affected by the heat transfer mechanism and the thermal conductivity of the cement-making materials (i.e., calcium oxide, CaO; silicon dioxide, SiO_2; alumina oxide, Al_2O_3; and ferric oxide, Fe_2O_3). When cement is made with conventional burning using fuel (e.g., oil and gas), the cement components are heated by an external heat source, and heat is transferred via conduction from the outside inward. Thermal properties, such as the specific heat, latent heat, and so on, regulate the kinetics of the heating process. However, cement-making materials have intrinsically low thermal conductivities and low heat transfer rates. Thus, they must be heated for long times until the inner portion of cement melts and sinters at a high temperature, to become a cement clinker nodule (at ~1450°C). To overcome this problem, microwave heating may be applied in the burning process of cement.

Unlike conventional heating, microwave heating principally occurs when the electric field of the microwave interacts with a material. Energy from the field is transferred to the molecular bonds of the materials. This energy transfer causes the bonds to vibrate and leads to heat dissipation within the material. Microwave heating is particularly efficient for heating dielectric materials (e.g., microwave heating was at least twice as effective as conventional methods for heating dielectric materials). Therefore, microwave heating should be useful in the cement and concrete industry, as cement-making materials exhibit excellent dielectric properties and should be able to absorb microwave energy very efficiently.

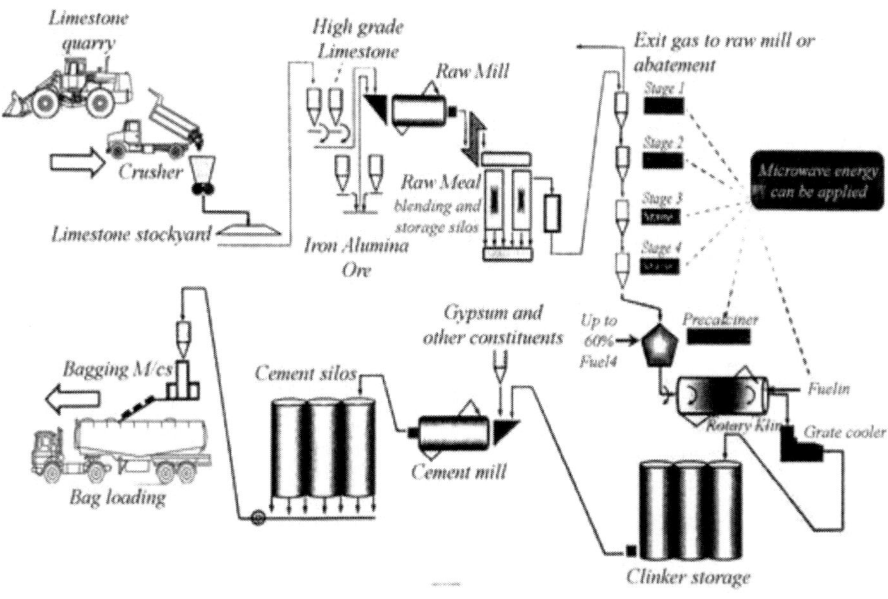

Figure 2.1. Overview of a conventional cement manufacturing process.

Furthermore, internal (volumetric) heating by microwave energy offers many benefits for the cement and concrete industries, including:

- Rapid heating rates and short processing times, which save energy and time;
- Deep penetration of the microwave energy (in cement and concrete materials, a microwave system operating at 2.45 GHz can typically penetrate several centimeters), which allows heat to be generated efficiently without directly contacting the cement constituents;
- Instantaneous and precise electronic control, which is convenient for designing and constructing heating systems;
- Unique and fine microstructural development, which permits better properties of produced cement; and
- Clean heating processes that do not generate secondary waste.

Based on International Telecommunication Union (ITU) regulations, the electromagnetic frequency band for non-telecommunication (e.g., industrial, scientific, and medical) purposes ranges from 6.765 MHz to 246 GHz. Microwaves are electromagnetic waves with frequencies ranging from 300 MHz to 300 GHz (wavelengths from 0.001 – 1.0 m). Microwave systems operate at average frequencies of 896 ± 10 MHz (in the UK), 915 ± 13 MHz (in North and South America), and 2450 ± 5 MHz (worldwide, except where 2375 ± 50 MHz is used), as assigned by the International Microwave Power Institute (IMPI). When used for industrial, scientific, and medical (ISM) purposes, microwave systems are most often operated in the range from 433.92 MHz to 40.00 GHz.

Many investigations and practices have recently reported using microwave heating for concrete applications, such as for accelerated curing, decommissioning and decontaminated surfaces, and drilling/melting. Moreover, microwave systems have been applied as high-performance nondestructive monitoring and surveying methods for cement and concrete structures.

This chapter offers a comprehensive and systematic review of the use of microwave energy in cement and concrete materials. This review systematically summarizes important results from previous experimental and numerical studies, covering such areas as the dielectric properties of cement and concrete, the mechanisms of microwave processing, the applications of microwave systems for cementitious materials, the health and safety of microwave processing, and future research and development (R&D) trends. Laboratory-based R&D and a practice-based scale innovation approach are discussed.

On the basis of findings in this review, some topics are recommended for further study and development, including the analysis of adaptive (time-dependent) dielectric properties, the coupling of heating mechanisms with the chemical reactions in cement and concrete, the design and construction of suitable microwave systems for use with cement and concrete, and the prediction of related phenomena (e.g., thermal runaway, as a highly regulated safety issue). Findings from this review can be used to guide future R&D for developing microwave energy applications in the cement and concrete field, in order to decrease energy consumption compared to other energy sources in heating/burning and to reduce the high temperatures needed for producing cement and curing concrete. These findings can also aid in the development of innovative and nondestructive microwave-based methods for monitoring the properties of cement and concrete products, for surveying deteriorated concrete structures, and for selecting suitable methods for concrete repair.

2.2. Dielectric Properties

Microwave heating involves the interactions between electromagnetic energy and the properties of dielectric material, especially dielectric propery. This property plays an important role to describe the behavior of concrete materials when subjected to microwave energy. This section describes theories of dielectric, measurement techniques of cement and concrete, and blended cement and concrete. Modeling work of calculation of cement is also explained.

2.2.1. Dielectric Theories

Cement and concrete are dielectric materials. Their intrinsic properties affect the way they interact with the electric and magnetic fields of the microwaves. Dielectric property can be characterized by two independent electromagnetic properties: the complex (electric) permittivity ε^* and the complex (magnetic) permeability μ^*. However, most common concrete materials are non-magnetic and yield permeability μ^* that is very close to the permeability of free space $(4\pi \times 10^{-7})(m{\cdot}kg/s^2.A^2)$. Thus, the complex (electric) permittivity ε^*, which comprises real and imaginary parts that can be defined as the relationship expressed in Eq. (2.1):

$$\varepsilon_r^* = \varepsilon_r' - \vec{j}\varepsilon_r'' \tag{2.1}$$

where ε_r' and ε_r'' are the real part and imaginary parts respectively, of the complex permittivity, and $\vec{j} = \sqrt{-1}$.

The real part of the relative complex permittivity, referred to as the dielectric constant, ε_r' , measures how much energy transferred from an external electric field is stored in a material. The imaginary one, ε_r'', measures how lossy a material is to an external electric field and is referred to as the relative loss factor. Besides an essential ratio that shows the energy lost (relative loss factor) to the energy stored (relative dielectric constant) in a material is given as loss tangent $tan\,\delta$, as shown in Eq. (2.2):

$$tan\,\delta = \varepsilon_r'' / \varepsilon_r' \tag{2.2}$$

The dielectric properties of the concrete including dielectric constant and loss factor are important parameters in microwave-dielectric material interactions. They are strongly dependent on moisture content, temperature, extent of hydration. Reports of prior investigations studies did not attempt to correlate experimental results with dielectric properties in order to develop a fundamental link to the field equations listed in Eq. (2.3), a task that must be accomplished in order to predict real-time changes during microwave curing. The internal distribution of free water in the sample is also an important consideration until the structure has been formed since this will decrease the efficiency of microwave heating.

$$\left.\begin{aligned}
\varepsilon(\omega) &= \varepsilon_\infty + \frac{\varepsilon_s - \varepsilon_\infty}{1+(i\omega\tau)^{1-\alpha}} - j\frac{\sigma(0)}{\omega\varepsilon_0} && \textit{(Maxwell - Wagner effect)} \\
\varepsilon(\omega) &= \varepsilon_\infty + \frac{\varepsilon_s - \varepsilon_\infty}{1+(2i\omega\tau)^{\beta} + i\omega\tau} - j\frac{\sigma(0)}{\omega\varepsilon_0} && \textit{(Double layer polarization)} \\
\varepsilon(\omega) &= \varepsilon_\infty + \frac{\varepsilon_s - \varepsilon_\infty}{1+(i\omega\tau)} - j\frac{\sigma(0)}{\omega\varepsilon_0} && \textit{(Debye effect)}
\end{aligned}\right\} \tag{2.3}$$

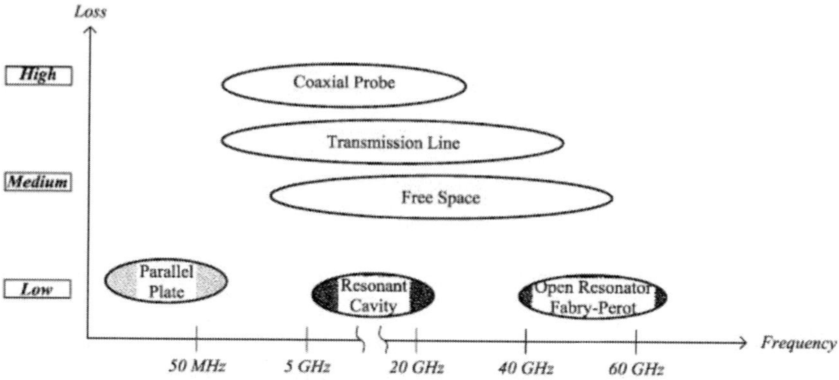

Figure 2.2. Summary of dielectric measuring techniques.

2.2.2. Measurement Techniques and Calculation

2.2.2.1. Measurement Techniques

Many techniques have been developed to understand the dielectric property (Table 2.2), including the transmission/reflection line, perturbation, free-space, and open-ended methods. Selection of the appropriate method depends, in part, on the operating frequency of the microwave system (Figure 2.2). For example, at the commonly utilized frequency of 2.45 ± 0.05 GHz, the dielectric property should be measured by transmission line. This method allows the measurement of specimens with a slab-shaped geometry placed across the cavity. However, this method is limited by the air-gap effect. The open-ended probe method could also be used, which involves measuring the phase and amplitude of the reflected signal at the end of an open-ended coaxial line inserted in a specimen. This method, too, has limitations, due to the reflection effect from the specimen.

Choices of the measurement equipment and sample holder design depend on several factors, including the dielectric materials to be characterized, extent of the research, as well as the equipment and resources that are available. Two types of dielectric analysers have recently been used: the scalar network analyser and the vector network analyser (VNA). The VNA is quite popular, very versatile, and useful for extensive studies; however, it is also very expensive. Although less expensive, scalar network analysers and impedance analysers are generally still too costly for many applications.

The HP 8510 is a VNA that has been used to measure the dielectric properties of cementitious materials, including cement paste, mortar, and concrete. It can measure the magnitude and phase characteristics of linear networks, such as filters, amplifiers, attenuators, and antennae. As with all network analysers, and as shown in Figure 2.3(a), the HP 8510 apparatus measures both reflection and transmission of electromagnetic waves at an interface. An incident signal generated by an RF source is compared with the signal transmission through the analyser or reflected from the wave input when passing through the waveguide (Figure 2.3 (b)). However, this method is strongly affected by the air-gap content, with high air content tending to decrease the dielectric property of the material. It is also influenced by the consistency of the cement and concrete materials. In particular, when materials behave like liquids, the settling of the material under its own weight can result in the non-uniform behavior of the whole specimen. Consequently, the measurement results are not representative of the whole body of the specimen.

A measurement using the transmission/reflection-line technique involves placing a sample in a section of the waveguide or coaxial line and measuring the two ports' complex scattering parameters with a VNA. Calibration must be carried out before carrying out the measurement, though. The technique involves measuring the reflected (S_{11}) and the transmitted signal (S_{21}). The relevant scattering parameters relate closely to the complex permittivity and permeability of the material by equations. The conversion of S-parameters to complex dielectric parameters is computed by solving the equations using a Nicholson-Ross-Weir (NRW) technique as shown in Figure 2.4.

Table 2.2. A summary of the technique for measuring dielectric properties

Technique	*Characteristics*	*Feature*	*Dielectric properties*
Transmission/ reflection line	The material under test (MUT) must be made into slab geometry. *Advantages:* • common coaxial lines and waveguide • used to determine both ε_r, μ_r of the MUT *Disadvantages:* • limit by the air-gap effect • limit to low accuracy	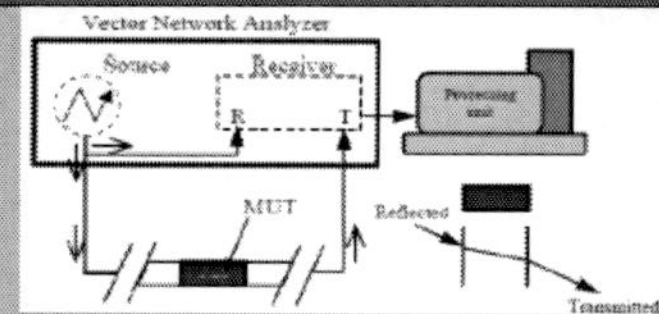	ε_r, μ_r
Perturbation (Cavity) (Resonant)	The measurement is made by placing a sample completely through the center of a waveguide that has been made into a cavity. *Advantages:* • simple and easy data reduction • accuracy *and* high temperature capability • suite to low dielectric loss materials *Disadvantages:* • need high frequency resolution vector network analysis (VNA) • limit to narrow band of frequency band only	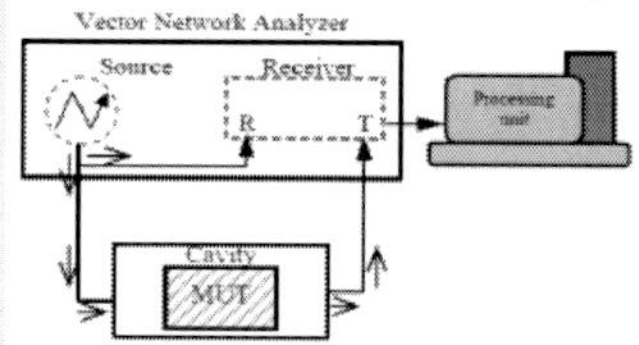	ε_r, μ_r
Free-space	The sample is placed between a transmitting antenna and a receiving antenna, and the attenuation and phase shift of the signal are measured. *Advantages:* • used for high frequency and allow non destructive measurement • measure MUT in hostile environment *Disadvantages:* • need large and flat MUT and diffraction effects at the edge of sample • limit to low accuracy	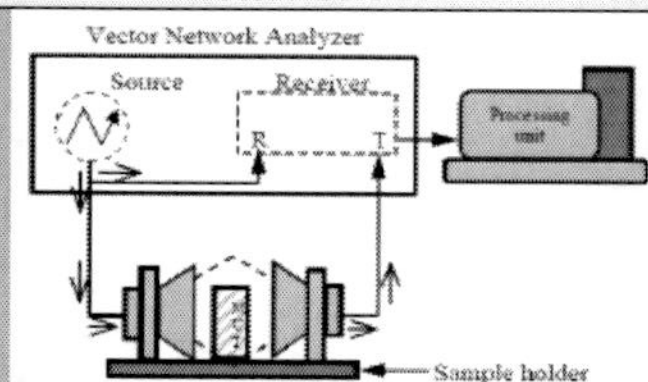	ε_r, μ_r
Open ended probe	The technique calculates the dielectric properties from the phase and amplitude of the reflected signal at the end of an open-ended coaxial line inserted into a sample to be measured. *Advantages:* • after calibration, can be routinely measured in a short time *Disadvantages:* • available for reflection measurement	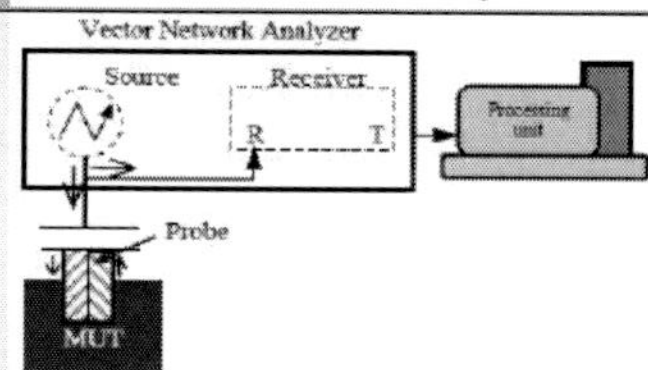	ε_r

Table 2.3. Dielectric constant (ε') and dielectric loss factor (ε'') of cementitious materials as function of temperature

Materials	*Dielectric constant ε (ε′); ε′(t)=at³ + bt² + ct + d*			
	a ±s.d.	*b ±s.d.*	*c ±s.d.*	*d ±s.d.*
Type 1 Portland cement	$-3.671\times10^{-6} \pm 1.276\times10^{-7}$	$3.529\times10^{-4} \pm 1.869\times10^{-5}$	$-2.014\times10^{-3} \pm 8.294\times10^{-3}$	$3.712 \pm 4.803\times10^{-1}$
Type 3 Portland cement	$-1.122\times10^{-6} \pm 1.599\times10^{-7}$	$4.021\times10^{-4} \pm 1.007\times10^{-5}$	$-1.476\times10^{-3} \pm 5.356\times10^{-4}$	$6.023 \pm 3.911\times10^{-1}$
Pulverized fuel ash	$3.333\times10^{-6} \pm 3.215\times10^{-7}$	$-6.667\times10^{-4} \pm 5.508\times10^{-5}$	$4.200\times10^{-2} \pm 2.859\times10^{-3}$	$4.138 \pm 3.109\times10^{-1}$
Silica fume	$-3.500\times10^{-7} \pm 7.071\times10^{-8}$	$5.500\times10^{-5} \pm 2.121\times10^{-5}$	$-3.350\times10^{-2} \pm 1.344\times10^{-3}$	$2.217 \pm 1.790\times10^{-1}$
Superplasticizer (polycarboxylic water-based) (42.5 % solid content by weight)	$-2.333\times10^{-5} \pm 8.083\times10^{-5}$	$4.767\times10^{-2} \pm 9.935\times10^{-3}$	$-3.260\times10^{-1} \pm 2.989\times10^{-2}$	$2.865 \pm 1.933\times10^{-1}$
Fine aggregate (River sand)	$-1.000\times10^{-7} \pm 7.000\times10^{-8}$	$-2.333\times10^{-5} \pm 1.079\times10^{-6}$	$2.333\times10^{-4} \pm 6.429\times10^{-5}$	$2.600 \pm 1.565\times10^{-1}$
Coarse aggregate (Crushed limestone rock	$-8.500\times10^{-7} \pm 1.626\times10^{-7}$	$2.600\times10^{-5} \pm 4.808\times10^{-6}$	$-1.950\times10^{-3} \pm 3.606\times10^{-4}$	$1.224 \pm 1.409\times10^{-1}$
Materials	*Dielectric loss factor (ε″); ε″(t)=et³ + ft² + gt + h*			
	e ±s.d.	*f ±s.d.*	*g ±s.d.*	*h ±s.d.*
Type 1 Portland cement	$3.333\times10^{-7} \pm 4.989\times10^{-8}$	$-7.667\times10^{-5} \pm 1.115\times10^{-6}$	$5.000\times10^{-3} \pm 7.874\times10^{-4}$	$1.565\times10^{-1} \pm 1.844\times10^{-2}$
Type 3 Portland cement	$7.987\times10^{-7} \pm 3.043\times10^{-8}$	$-3.008\times10^{-5} \pm 9.453\times10^{-6}$	$6.225\times10^{-3} \pm 2.114\times10^{-4}$	$3.922\times10^{-1} \pm 1.042\times10^{-2}$
Pulverized fuel ash	$-1.800\times10^{-7} \pm 3.111\times10^{-8}$	$2.650\times10^{-5} \pm 4.738\times10^{-6}$	$-1.500\times10^{-3} \pm 2.687\times10^{-4}$	$2.763\times10^{-1} \pm 7.092\times10^{-2}$
Silica fume	$-3.667\times10^{-7} \pm 2.082\times10^{-8}$	$6.333\times10^{-5} \pm 3.512\times10^{-6}$	$-3.733\times10^{-3} \pm 1.850\times10^{-4}$	$2.223\times10^{-1} \pm 2.045\times10^{-2}$
Superplasticizer (polycarboxylic water-based) (42.5 % solid content by weight)	$-2.000\times10^{-5} \pm 1.414\times10^{-6}$	$3.400\times10^{-3} \pm 1.273\times10^{-3}$	$-1.664\times10^{-1} \pm 5.155\times10^{-2}$	9.635 ± 1.025
Fine aggregate (river sand)	$7.000\times10^{-8} \pm 1.224\times10^{-9}$	$-2.000\times10^{-5} \pm 2.633\times10^{-6}$	$1.000\times10^{-3} \pm 1.467\times10^{-4}$	$2.645\times10^{-1} \pm 6.029\times10^{-2}$
Coarse aggregate (crushed limestone rock)	$7.000\times10^{-8} \pm 1.453\times10^{-9}$	$-2.000\times10^{-5} \pm 8.201\times10^{-7}$	$1.000\times10^{-4} \pm 4.798\times10^{-5}$	$5.680\times10^{-2} \pm 4.334\times10^{-3}$

Remark: s.d. = Standard Deviation.

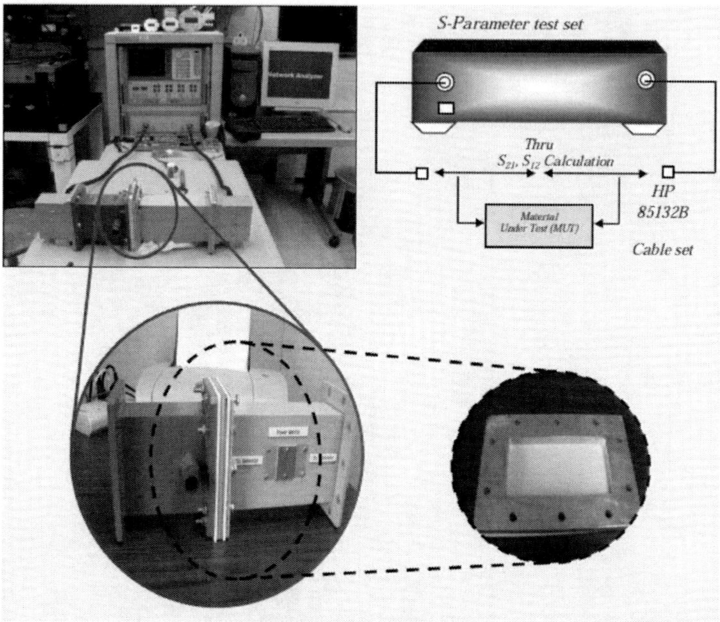

Figure 2.3. (a) Network analyzer, (b) Schematic of S-parameter test.

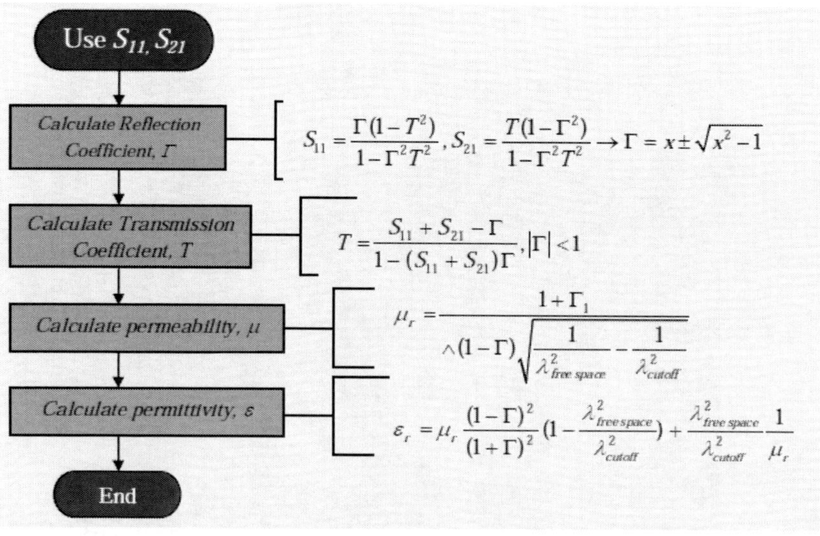

Figure 2.4. Nicholson–Ross–Weir conversion process for calculating dielectric permittivity.

2.2.3. Measurement of Cement and Concrete

There have been several investigations of the dielectric properties of cement and concrete materials, covering their measurement methods, their behaviors, and the factors influencing their variations. Initially, microwave radiation was used to determine the evaporable moisture (free-water) content in building materials and structural elements by monitoring the absorption level of microwave. Later, Hasted and *Shah* used the Robert and Von-Hippel method, a time-domain reflectometry-based measurement method, to study standing-wave behavior within a waveguide at 10, 3.3, and 1.25 cm. Wittmann and Schlude studied microwave absorption at 9 to 11 GHz by hardened cement paste, using a free-wave method with a specimen of specified diameter and height. They consistently agreed that the real and imaginary parts of the complex permittivity of cement could be determined as a function of the moisture content, temperature, and hydration duration of the cement, when using the transmission/reflection line method or a cavity-resonant method.

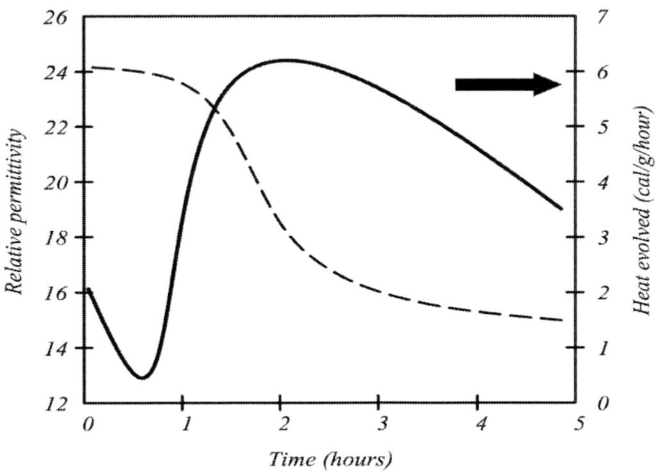

Figure 2.5. Correspondent conductivity and heat of hydration for type 1 0.4-w/c cement paste.

The permittivity of a concrete mixture is influenced by the w/c ratio under various free-water volume absorption conditions. Variations in permittivity, due to the effects of the w/c ratio, cement type, and superplasticizer use, are closely related to changes during the four stages of cement hydration. The ion concentration is correlated with the mixing of the cement and water particles. When the cement first contacts with water, the ion concentration increases to a relatively high level. It remains constant during the dormant period, decreases rapidly during the hydration reaction, and continues to decrease during the acceleratory period, as shown in Figure 2.5. Some typical dielectric constant (ε') and dielectric loss factor (ε'') values of cementitious materials as a function of temperature are shown in Table 2.3. Many microwave-based methods are available for monitoring variations in the dielectric properties during the hydration process. Microwave heating units are commonly operated at 2.45 GHz, which is a suitable frequency for monitoring concrete mixtures using a custom-designed parallel-plate setup with a network analyzer.

The dielectric properties of cement and concrete materials change drastically and continually in the early stages of concrete formation (i.e., from the time when cement particles first contact with water molecules until the final setting occurs). Parameters that affect the dielectric property during this process should be considered. For example, the measurement method should be able to represent the whole specimen. The method should not be affected by external moisture or temperature, nor should it be set in the system to be measured, as it is in an adiabatic system. In addition, the strong sensitivity of cement and concrete materials to each microwave range needs to be considered.

2.2.4. Measurement of Blended Cement and Concrete

In addition to conventional cement, researchers have investigated the dielectric properties of cement-pozzolan mixtures, as well as blended cement and concrete. For example, using a perturbation method with microwave systems operated at 8.2 to 12.4 GHz during the early hydration reaction, researchers found that the dielectric constant increased and the induction period decreased with an increasing percentage of slag in cement paste. A breakthrough in the dielectric measurement of materials containing high-performance cement-pozzolanic materials indicated that the relative dielectric constant at a microwave frequency of 10 kHz ± 1 MHz decreased from 29 to 21 with the addition of increasing silica fume content.

Later, Levita et al. studied the electrical properties of liquefied type 1 52.5 R Portland cement (European standard ENV- 197/1) containing microsilica fume during the early stage of hydration. These authors made measurements at ambient temperature for up to 40 h, in the frequency interval from 30 Hz to 200 kHz. The electrical properties of the paste mixtures had major variations in the first 40 h of hydration. They reported the values of ε' and ε'' as a function of frequency. At low frequencies, the dielectric constant was very large (~10^8) because of an induced dielectric amplification effect, which was previously described by Coverdale et al. and Ding et al. This effect involved a decrease of the dielectric permittivity with the progression of hydration, due to an increasing amount of calcium-silicate hydrate (C-S-H) or the conversion of free water to combined water in the C-S-H molecule. The permittivity varied depending on the chemical composition of the constituents, the amounts of hydration product phases, and local topological variations. At high frequencies, the dielectric amplification effect was significantly reduced.

Dielectric properties of fresh concrete compositions during the early stage of hydration were also investigated. The dielectric constant varied with time between 30 and 60 min after cement was mixed with water. These results were consistent with the findings of Paul and Stephen, who used frequencies from 100 Hz to 7 MHz, Ping and Beudoin, who used frequencies from 1 to 1000 MHz, and Youssef et al., who used frequencies from 10 Hz to 1 GHz.

The electromagnetic characteristics and absorbing properties of fly ash were studied for cement composite in which fly ash was used as a cement replacement. The ε' was almost constant (~3.15) in the whole frequency range from 2 to 8 GHz, whereas ε'' showed obvious changes in the range from 0.03 to 0.1 GHz. Several peaks in ε'' were obtained at fixed positions. The work of Makul et al. comprehensively investigated the behaviors of various blended cement and concrete mixtures during a 24-h first-hydration period at 2.45 ± 0.05 GHz. The dielectric permittivity was relatively high and remained constant during the dormant period. Thereafter, it decreased rapidly when the hydration reaction resumed, and continued to decrease during the acceleratory period, consistent with the works of Moukwa et al. and Zhang et al.

Apart from experimental investigations, Hager and Domszy tried to fit the experimental results of hydrating cement paste to a model over the frequency range of 10 kHz to 8 GHz, from initial mixing through several weeks of cure. The model included the following terms: (1) Cole-Davidson relaxation near 1 MHz, (2) Debye relaxation near 100 MHz, (3) free-water relaxation near 10 GHz, and (4) ion conductivity and electrode polarization. The first two terms initially increased, and all four terms decreased with advancing cure. This work was the first to present long-term changes in the dielectric properties of cement, and it is applicable for pozzolanic cements with long-term reactions. The classical effective permittivity theory described by Prasad and Prasad indicates that low-permittivity inclusions can decrease the effective permittivity of composites. In that study, the Cuming equation had the highest degree of acceptability (errors < ±15%) in all cases.

Some limitations of the aforementioned investigations should be noted. First, the dielectric constant fluctuates during the early stage of the hydration reaction. Second, other equations may be more optimal than the Cuming equation, such as the Maxwell-Wagner, Webmann, Skipetrov, and modified Cule-Torquato equations, depending on the sensitivity of the moisture (water) content of the cementitious materials.

2.2.5. Modeling Work

In addition to the experimental studies described above, theoretical-based analyses of the dielectric properties of cementitious materials have been performed. For example, finite difference-time domain (FD-TD) modeling has been used to visualize the propagation of electromagnetic fields in concrete (as a dielectric medium) and to predict electromagnetic phenomena associated with concrete targets for nondestructive testing purposes. Klysz et al. proposed a numerical FD-TD model of a ground-penetrating radar-coupled antenna to evaluate the dielectric properties of concrete. Using their method, the permittivity and conductivity of different concretes at different degrees of saturation were in agreement with those provided by Soutsos et al., who used a coaxial transmission line system. These results were later validated in model experiments by an open-ended coaxial probe method over a microwave frequency range from 0.1 to 20 GHz. The results suggested that the outer geometry of the specimens (i.e., shape and size) affected the ability to detect the targets and inclusions.

A survey of proposed mathematical models and validations indicated that the model needs to be validated with various concrete mixtures with different internal moisture contents because the latter factor strongly affects the dielectric constant and loss factor of concrete. The dielectric parameters (ε' and ε'') are not linked to the porosity; rather, all of the concrete pores must be filled with moisture in both the saturated and partially saturated states, especially in the early stage of hydration. Therefore, additional w/c ratios of concrete will need to be investigated in detail.

2.3. MICROWAVE SYSTEMS

This section presents the components and function of existing microwave systems, as well as their practical use for different applications. Conventionally a suitable microwave system is required to generate and introduce microwave energy to material (dielectric) to be heated. A typical microwave system is shown in Figure 2.6. Most microwave-heating systems consist of three basic components: microwave power generator (usually magnetron), wave guide, and microwave chamber or cavity where the specimen is to be processed and supplementary components e.g., control system, tuner, circulator and so on. In detail, most importantly, microwave power generator connecting to power supply unit: its purpose is to convert the alternating current (a.c.) of 50 or 60 Hz to a desired high frequency microwave radiation. This is achieved by a device called magnetron. Magnetron must also be shielded to prevent the leakage radiation of harmonics, as well as for safety. Waveguide is an important component and it transmits microwave energy through free space and delivers to the specimen. The waveguide is any structure that directs the propagation of energy in the form of an electromagnetic wave along a predetermined path. Waveguides may assume many different physical forms depending up on a large extent on the frequency band. Last is microwave chamber/cavity that obeys with the laws of electromagnetic—transmit, absorb, reflect microwave energy.

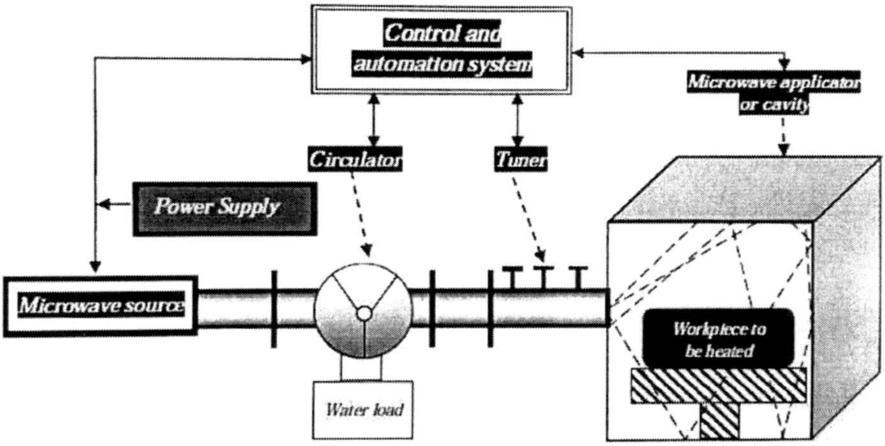

Figure 2.6. Microwave system diagram.

Practically, microwave system can be categorized into three subsystems depending on type of applicator associated with supplementary equipment or system as follows:

- Traveling-wave applicators in which a traveling wave is produced by matching the microwave source to a heated load using rectangular and cylindrical waveguide.
- Near-field applicators—this is a resonant open-ended waveguide or cavity attached directly at the end of the waveguide, at which the energy is fed through slots into a cavity.
- Resonant applicators—this is these types can be classified into single and multimode resonators as shown in Figure 2.7 (single mode) and Figure 2.8 (multimode), respectively.
  - For single mode resonant cavity, the standing wave applicator is attached to the end of the waveguide microwave power-feed line. Microwave energy was generated by a magnetron and transmitted directly along the propagation direction ($+z$) of a rectangular waveguide

toward a water load situated at the end of the waveguide to ensure that a minimal amount of microwave energy would be reflected back to the sample. A warming water load was circulated through the cooling tower in order to reduce the temperature in the water load system.

- For multimode applicators, Figure 2.8 (a), the microwave cavity was rectangular shape with exterior dimensions of 900 mm × 450 mm × 2700 mm. The drier operates at a frequency of 2.45 GHz with maximum working temperature of 180 °C. The microwave power was generated by means of 12 compressed air-cooled magnetrons. The maximum microwave capacity was 9.6 kW with a frequency of 2.45 GHz. The power setting could be adjusted individually in 800 W/cteps. In the continuous processing equipment, two open ends were essential, through which the material to be heated up on the belt conveyer was put in and taken out. In this equipment, leakage of microwaves was prevented by the countermeasure in double with a combination of mechanical blocking filter (corrugate choke) and microwave absorber zone filter was provided at each of the open ends. The multiple magnetrons (12 units) were installed in an asymmetrical position on the rectangular cavity (Figure 2.8(b)). The microwave power was then directly supplied into the drier by using waveguides. An infrared thermometer (located at the opening ends) was used to measure the temperature of the specimens (accurate to ± 0.5°C). An example is high temperature microwave system. It can generate microwave energy at a frequency of 2.45 GHz and maximum power of 6.0 kW with a multi-mode system. This results in a temperature of 2000 °C being induced in the material.

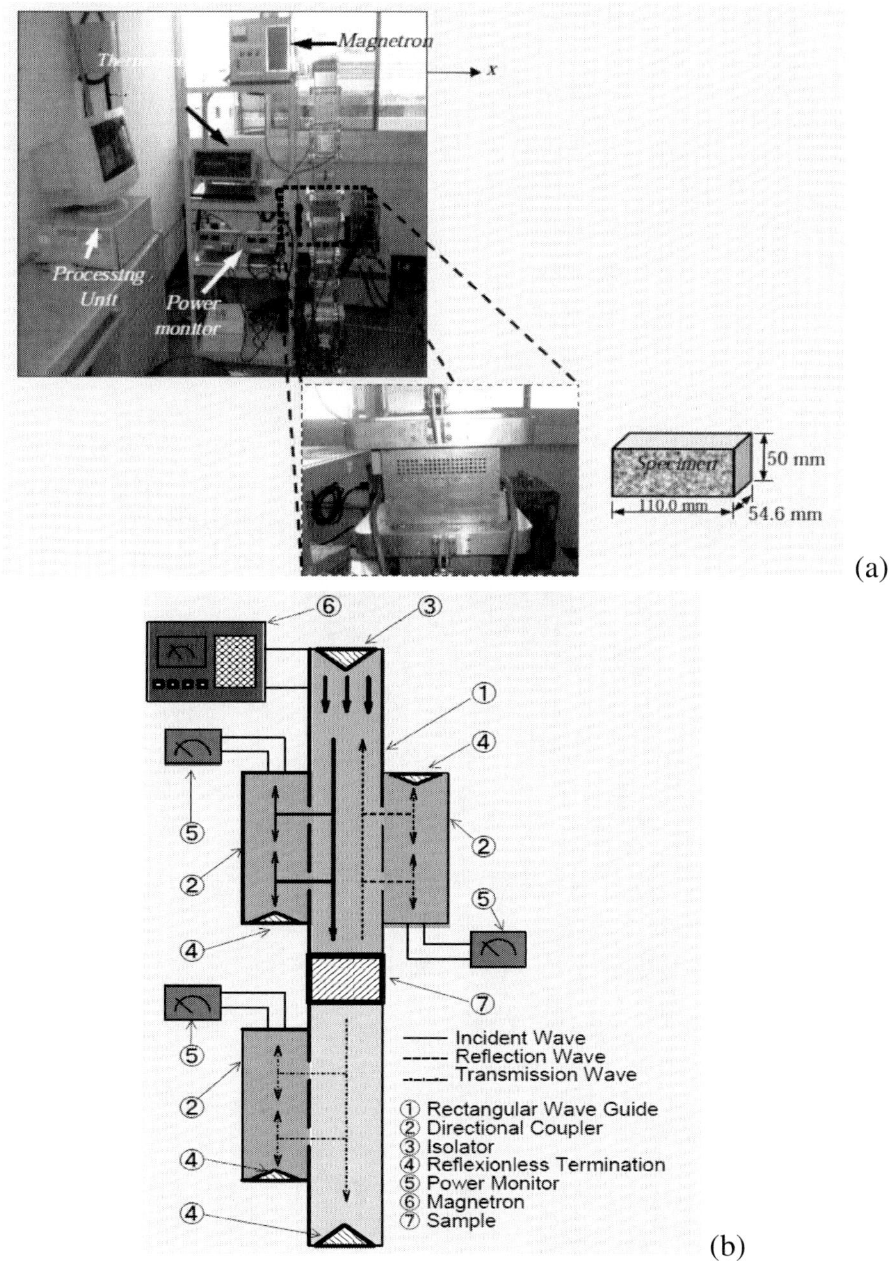

Figure 2.7. (a) Experimental set up, (b) schematic showing direction of microwave components (Incident wave, reflected wave, and transmission wave).

Cement is a complicated mixture of calcareous-, silica-, and alumina-based minerals with specific compositions. The manufacturing process consists of grinding and mixing the raw materials, and then roasting at temperatures up to 1450°C. During this process, a granular clinker is formed, which is composed of tricalcium silicate ($3CaO \cdot SiO_2$ or C_3S), dicalcium silicate ($2CaO \cdot SiO_2$ or C_2S), tricalcium aluminate ($3CaO \cdot Al_2O_3$), and tetracalcium aluminoferrite ($4CaO \cdot Al_2O_3 \cdot Fe_2O_3$). To generate commercial Portland powder, clinkers are ground to a fine powder and 4 – 5% calcium sulfate is added. The resulting product is ready to be reacted with water.

In the following sections, various applications of microwave energy to cement are presented, covering the use of microwave energy to synthesize cement and to improve cementitious properties.

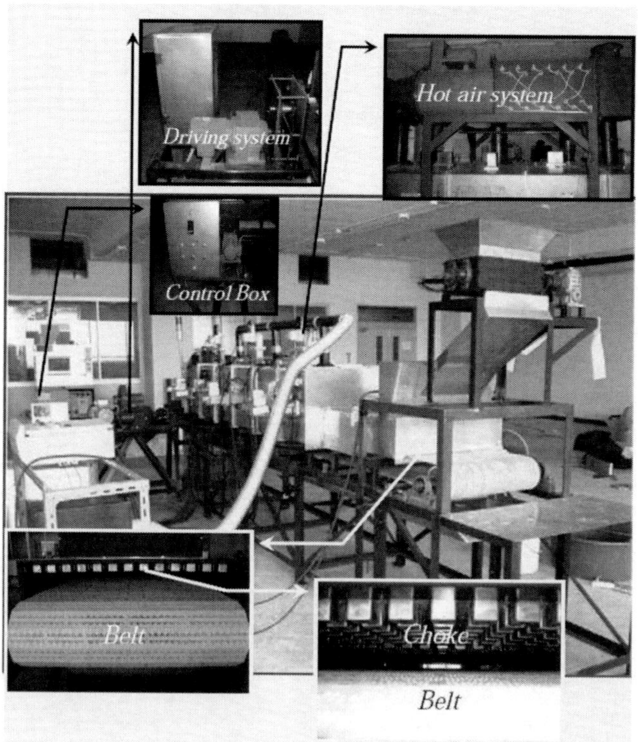

(a)

Figure 2.8. (Continued).

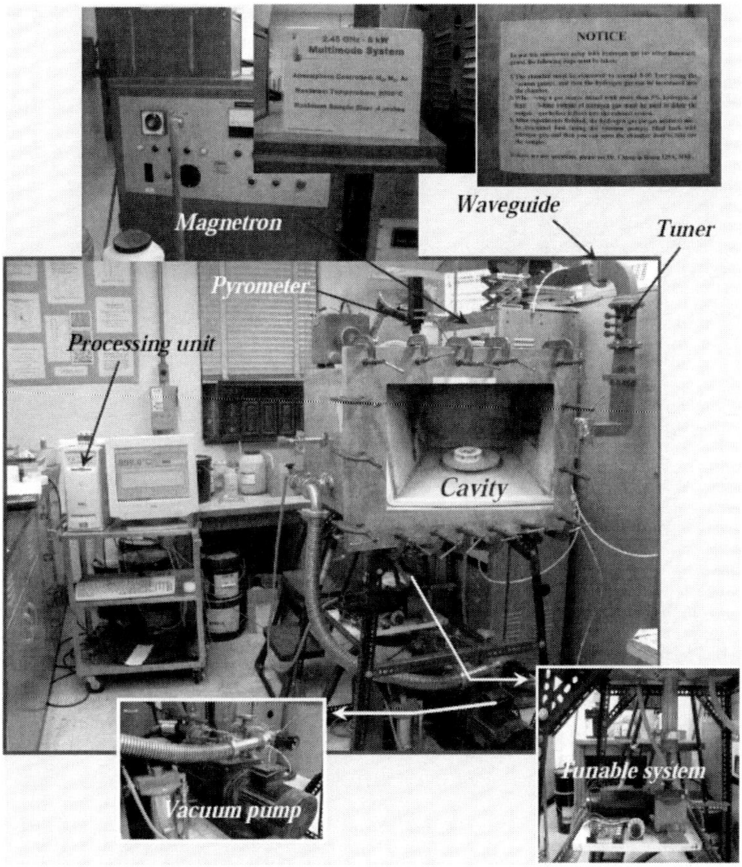

(b)

Figure 2.8. Microwave system typed multimode resonant cavity: (a) A combined multi-feed microwave-hot air heating with continuous belt system (Low temperature multi-mode microwave system), (b) High temperature multi-mode microwave system.

2.4. Applications of Microwave to Cement

2.4.1. Cement Synthesis

Microwaves have been utilized at 2.45 GHz to assist in the clinkerization of cement nodules, producing clinkers with similar characteristics to those of conventionally produced industrial cement clinkers. During microwave heating at a maximum temperature of 1450°C, the dielectric losses of different cement constituents were adequate to preserve the heating rate and ensure clinkerization, as confirmed by Sutton. In a later work, microwave energy at 2.45 GHz was capable of synthesizing regular and colored cement clinkers. Clinkering temperatures required by microwave processing were about 100°C less than those required for fast-heating by the electric furnace method. Another microwave sintering method involved heating the sample by electric heating to 1500°C, and then using microwave-assisted heating for 30 to 60 min to induce C_3S formation. Ion oxides played a very important role in C_3S formation in conventional sintering.

Microwave heating has been used to sinter Class F fly ash at temperatures from 800 to 1000 °C for 10 to 20 min. Samples sintered by microwave were denser and had better mechanical properties than samples sintered by conventional processes at the same sintering temperature and for the same time. The sintered cement nodules were mainly composed of mullite ($Al_6Si_2O_{13}$). At the same heating rate during clinker formation, microwave-assisted sintering improved the formation of C_3S compared to conventional sintering methods.

In addition to the use of only microwave heating for cement sintering, a combined electric and microwave clinkering method has been applied for Portland cement. After raw materials had been heated by electric energy to 1000 – 1200 °C, they were sintered by microwave energy for 1 to 2 min. The sintering time was reduced by using higher temperature samples in the microwave-sintering step. The same combined method was also applied to make sulphoaluminate belite (SAB) cement clinkers from a raw sulfoaluminate precursor. The combined use of electric heating and microwave processing significantly accelerated the clinkering reaction. For samples that were heated at 1300 °C for 1 h, the free CaO of the obtained clinkers was as high as 1.03 – 4.78%. When samples of the same composition were heated by microwave alone for 25 min, the $CaCO_3$ in the raw materials did not completely decompose.

2.4.2. Improving Cementitious Properties

Microwave can be utilized to improve the cementitious properties of pozzolan materials. For example, rapid sintering of rice husk ash (RHA) at 800 to 1200°C by microwave at 2.45 GHz enhanced the cementitious properties of the cementing material. The sintered and ground RHA contained the cristolbalite phase. The strength index of 28-day-old microwave-sintered RHA paste sintered at 800°C was increased by 30% compared to the index before sintering.

2.5. Applications of Microwave to Concrete

This section presents various applications of microwave energy to concrete, including microwave-accelerated curing, the decommissioning of decontaminated concrete, nondestructive monitoring, and the drilling/melting of concrete.

2.5.1. Acceleration of Curing

After cement grains are mixed with sufficient water, various reactions occur that result in hydration product formation (e.g., calcium silicate hydrate gel, C_x-S_y-H_z) (Eq. (4)). Due to the nature of the cement grains and the reaction kinetics, the strength of the concrete continually increases over time. Immediately after the cement and water are mixed, the $CaO{\cdot}Al_2O_3$ phase reacts with water to form an aluminate-rich material, and calcium sulfate hemihydrate is converted to its dihydrate form. Both of these reactions contribute to a brief exothermic peak. Subsequently, the gel reacts with sulfate in solution to form ettringite ($3CaO{\cdot}Al_2O_3$) and, thereafter, enters a dormant period .

At the end of the dormant period, $3CaO.SiO_2$ and $2CaO.SiO_2$ in the cement begin to hydrate, forming calcium silicate hydrate (CaO_x-SiO_y-H_2O_z) and calcium hydroxide ($Ca(OH)_2$). This step corresponds to the main period of hydration, when much of the increase in strength occurs. The cement grains react from the surface inwards, and unreacted grains become fewer and eventually are consumed. The period of maximum heat evolution occurs between 6 and 24 h after mixing, followed by a gradual decrease approximately over 1 month period. The sequence of reactions is shown below:

$$\left.\begin{array}{l}
3CaO\cdot SiO_2+H_2O \rightarrow CaO_x\text{-}SiO_{2y}\text{-}H_2O_z+Ca(OH)_2+\Delta Heat\ (500\ J/g) \\
2CaO\cdot SiO_2+H_2O \rightarrow CaO_x\text{-}SiO_{2y}\text{-}H_2O_z+Ca(OH)_2+\Delta Heat\ (260\ J/g) \\
2(3CaO\cdot Al_2O_3)+18H_2O \rightarrow 2CaO\cdot Al_2O_3.8H_2O+4CaO\cdot Al_2O_3\cdot 10H_2O+\Delta Heat\ (866\ J/g) \\
2(3CaO\cdot Al_2O_3)+32H_2O+3(Ca^{2+}_{(aq)}+SO_4^{2-}{}_{(aq)}) \rightarrow \\
\quad \rightarrow CaO\cdot Al_2O_3\cdot 3SO_3\cdot 32H_2O+\Delta Heat\ (420\ J/g) \\
6CaO\cdot Al_2O_3\cdot 3SO_3\cdot 32H_2O+23CaO\cdot Al_2O_3 \rightarrow 3(4CaO\cdot Al_2O_3\cdot SO_3\cdot 12H_2O)
\end{array}\right\} \quad (2.4)$$

Although cement can react with water under conditions of room temperature and pressure, such cement will require more time to reach an adequate strength for industrial uses. Consequently, various curing methods are used to accelerate strength development in cement-based materials during the early-age phase.

Strength is a crucial factor in structural design. High early strength can benefit concrete production by reducing the construction time, labor, formwork, energy, and environmental impact. Early strength development is strongly influenced by several factors, including the water/cementitious material ratio, cement type and content, age of the paste, and curing method. For example, use of a lower water/ cementitious materials ratio (w/c) makes a stronger concrete.

Many methods are available for increasing the early strength of concrete. Curing involves maintaining favorable moisture and temperature conditions to ensure that the degree of hydration is sufficient to reduce the porosity to a level at which the desired properties (e.g., strength and flexibility) can be attained. Curing affects the rate of strength development. Accelerated gains in strength may be achieved by supplying heat and moisture to the concrete. For example, curing with steam at atmospheric pressure (live steam curing) or at high pressure in an autoclave (high-pressure steam curing) is advantageous when high early strength is necessary or when additional heat is required to accomplish hydration (e.g., in cold weather).

Conventional curing methods have many limitations. For instance, when water curing is used, long periods are needed to reach the required strength. With steam curing, non-uniform temperature distributions can occur, resulting in non-homogeneous curing. When admixtures are employed, durability problems may occur. Thermal curing methods are especially susceptible to negative effects on durability in both the early and long-term stages. For example, high-temperature curing at atmospheric pressure was shown to result in reduced long-term strength and durability problems. Although the reasons for these problems are not wholly understood, they might include an increase in microcracking and delayed ettringite formation. As mentioned in Verbeck and Helmuth, rapidly accelerated hydration during curing can lead to the encapsulation of the anhydrous cement grains by a layer of product with low porosity, which further slows the hydration. Hydration products act as insulating (dielectric) materials that transfer heat at a low, non-uniform rate, resulting in poor properties after curing.

Moving the heat source from the outside surface of a heated cementitious material to the internal structure (e.g., through the interaction between the microwave field and the heated material) can result in a violent cement-water interaction and volumetric heat generation. This concept serves as the basis of one method of accelerated curing (microwave curing). The Portland cement-water system inherently reacts to a complex multicomponent system and repletion-coupled temperature- and composition-sensitive hydration products. Therefore, to develop the accelerated curing method further,

Portland cement-based materials subjected to high-temperature curing by microwave energy will need to be characterized from a microstructural perspective.

Given the adverse short- and long-term effects that thermal-curing methods have on the properties of concrete, a crucial question is whether microwave heat can be applied in the concrete industry. Theoretically, this application is possible because the materials used to make concrete (e.g., hydraulic Portland cement, aggregates, water, and admixtures) are dielectric (i.e., they can absorb microwave energy). In particular, the relative dielectric constant (ε_r') and loss tangent $(tan\,\delta)$ values for water are higher than those of the other components of cement. As a result, when the electric part of the electromagnetic field $(\bar{E})$ interacts with the constituents of concrete, energy is transferred from the electric and magnetic fields and converted (through ion conduction and polar rotation) to the molecular bonds of the water, which is part of the microwave-irradiated material. This mechanism causes the bonds to vibrate. Energy is dissipated as heat and transferred within the concrete, increasing the temperature and accelerating hydration. Free water molecules in the capillary pores of the concrete are quickly removed from the internal concrete structure before setting. Thus, plastic shrinkage occurs, which leads to collapse of the capillary pores and densification of the microstructure.

Studies on the use of microwave-assisted curing have numerous aspects. A paper by Watson was the first to extend a microwave dehydration method to concrete. Use of microwave energy accelerated strength development in Portland cement mortar without decreasing the long-term strength of the concrete. The strength of the concrete specimens began to decrease when the final w/c ratio was less than 0.40. In another study, microwave-assisted curing during the first 24 h reduced the induction period of hydration. Key parameters for microwave curing include the microwave power, application time, duration of microwave heating, thermal runaway, and overheating within the sample. When optimal microwave power was applied in a discretized fashion through feedback temperature control, very good results were obtained for concrete. The work of Sohn and Johnson obtained the optimal microwave-curing of type 1 Portland cement mortars at 40 and 60°C, whereas other temperatures were incompatible with curing.

Microwave curing appeared to damage the freeze-thaw durability of high w/c concrete, but not low w/c concrete. In terms of the microstructural characteristics of cement-based materials subjected to microwave energy, microwave-cured paste demonstrates some ettringite. In two studies, the ranges of the Si/Ca and Al/Ca ratios in the autoclave-cured paste were 0.039 to 0.052 and 0.207 to 0.234, respectively, whereas the ratios in the microwave-cured paste ranged from 0.079 to 0.091.

Apart from conventional concrete, microwave heating can also be used to cure metakaolin-blended cements. For instance, the pozzolanic activity of metakaolin was improved under microwave curing. Microwave-cured 15% metakaolin was free of $CaOH_2$, whereas 30% or 40% metakaolin was needed for similar outcomes in room-temperature curing. Concrete samples containing 10% silica fume responded well to microwave curing. Microwave heating for 40 min appeared to be the optimal curing time.

Microwave energy can also be used in association with a vacuum system, in order to dewater or dehydrate the concrete. The optimal microwave curing time has been shown to be 45 min at a maximum temperature of 60°C, with a final w/c ratio of around 0.38. In this process, microwave curing decreases the amount of free water while increasing the hydration rate of cement, resulting in a compacted/less porous concrete. When microwave energy was applied with a continuous belt drier consisting of 14 compressed air-cooled magnetrons of 800 W each (maximum of 11.2 kW), the microwave energy accelerated the early-age compressive strength of the cement paste, but did not affect the later-age strength. Microwave curing reduced the energy consumption and curing time.

Practically, microwave-assisted curing has been shown to be a useful technique for construction applications and for repairing damaged concrete pavements. Leung and Pheeraphan compared the early-age and long-term compressive strengths of concrete specimens developed from microwave-cured type 3 Portland cement to those of commercially available rapid-hardening concrete containing accelerating admixtures. The compressive strength of microwave-cured concrete (w/c ratio of 0.55) that was cured for 45 min increased to 19.2 MPa at 4.5 h. Studies based on research by the Commonwealth Scientific and Industrial Research Organisation (CSIRO) of Australia found that the use of microwave curing cycles of less than 6 h permitted adequate strength for formwork removal and wire-strand prestressing, without impairing the concrete quality.

Microwave heating offers efficient production and high product performance for precast concrete. Heat curing impacts the performance (i.e., thermal response, early-age strength, and durability) of a selected range of concrete materials for bridge construction. Microwave heating produces considerably lower temperature gradients compared to steam heating. For example, when rapid microwave curing cycles of less than 6 h were used, compressive strengths of up to 25 MPa were achieved in high-quality precast concrete. Replicating the bulk heating rate with microwave heating did not cause any deterioration in the near-surface quality, unlike in conventional steam heating.

Several inventions have been developed for the application of microwave-accelerated curing. Mario and Sergio developed a process and device for accelerating the drying of cement mixtures, such as those containing prestressed and non-prestressed concrete components. Ludewig and Steinbach invented a method and device for removing water from buildings, parts of buildings, and/or fixed structures, including those outside buildings, for construction purposes. Birch-Rasmussen et al. invented a process for curing and drying wet reinforced concrete. In their process, the concrete is subjected to short-term electromagnetic heating (for seconds to a few min), which increases the temperature of the concrete from 10 – 25°C to 50 – 80°C. Furthermore, Engelbrecht and Soeren proposed a process for curing and drying concrete by electromagnetic radiation, and a multistep process for accelerating the curing and drying of reinforced concrete products. In their proposals, the cast concrete elements are consecutively subjected to an inductive alternating magnetic field, high-frequency electromagnetic radiation, and electromagnetic microwave radiation.

Although the initial cost of the microwave equipment might be expensive, microwave curing can reduce the amount of cement used in concrete and can accelerate early-age concrete curing in various applications. However, this curing method could have negative impacts on the performance (i.e., thermal response, early-age strength, and durability) of some concrete mixtures. Therefore, a suitable microwave system and appropriate curing procedures need to be determined.

2.5.2. Decontamination and Decommissioning of Concrete

Microwave energy has been applied to the decontamination of concrete surfaces, including radioactively contaminated surfaces, and concrete layers, in the case of biological shields. When a targeted dielectric material is irradiated by microwave energy, the energy is transferred from the electric field to the molecular bonds of the concrete, leading to vibration of molecules within the bond. Due to the non-uniform temperature distribution under the concrete surface, water beneath the surface evaporates, generating a high vapor pressure on the surface. When the accumulated pressure exceeds the tensile strength (maximum tensile stress) of the concrete, the concrete surface will spall off the bulk concrete.

A series of studies investigated the decontamination of concrete by microwave technology under operating frequencies of 0.896, 2.45, 10.6, and 18.0 GHz. For microwave frequencies at or above 10.6 GHz, the microwave power dissipation was shifted towards the front surface of the concrete. This finding was verified in theoretical studies by Lagos et al. When higher microwave frequencies were used, less power was needed to increase the temperature or thermal stress to the same value in the same period of time compared to experiments under lower frequencies. These results are consistent with the findings of Bažant and Zi, who used microwave to spall off a thin layer of concrete surface that had been contaminated with radionuclide.

More investigations are needed into the use of microwave energy to spall off targeted concrete surfaces for decontamination. The decontamination of steel-reinforced concrete needs to be studied carefully because steel can reflect microwave energy, causing overheating and damage of the bulk concrete. Moreover, the decontamination of the concrete surface to make a break at a specific point or area requires careful analysis. Specifically, the design of an apparatus or machine to control the breaking region should be carefully considered. Finally, the safety of the worker should be taken into consideration.

2.5.3. Nondestructive Monitoring

The early strength and accelerated curing parameters of microwave-cured concrete have been used to predict the 28-day compressive strength of concrete. The 7-day strength of concrete made with rapid-hardening Portland cement and the 28-day strength of concrete made with ordinary Portland cement were predicted from the accelerated strength of concrete by microwave energy. Prediction results were within 15% agreement with actual test results.

Near-field microwave-sensing techniques can be used to determine the hydration state of cement paste or concrete. For example, microwave energy has been applied to determine the natural moisture content, percent absorption, and bulk specific gravity values of aggregates, to study the hydration kinetics, to observe the hydraulic reaction of fresh pastes of C_3S for the first 30 h of hydration, and to determine the water content of fresh concrete. The compressive strength of hardened concrete can be estimated from the apparent activation energy using a microwave technique. This technique determines the strength in a nondestructive manner, without contacting the sample. In addition, microwave has been used to monitor the residual moisture content in the early hydration stages of cement paste. The residual moisture content decreased in rough proportion to the stage of the hardening process.

In the case of reinforced or prestressed concrete, microwave can indicate the presence of water, which is important for the diagnosis of concrete. The propagation of electromagnetic waves in concrete is controlled by the electromagnetic properties of the material itself, which are mainly influenced by the presence of water. Concrete-independent linear relationships were observed between the water volume and the propagation velocity of the direct wave and its attenuation.

For damaged concrete, microwave can be used to distinguish mortars containing alkali-silica reactive (ASR) aggregate from nonreactive aggregate. Variations in measurements were linked to the production of ASR gel and its tendency to attract free water from its environment. ASR gel in cement-based materials can also be detected by material characterization methods. In order to utilize microwave to measure the concrete properties, the status of free and fixed water must be considered. The temperature of free water can increase due to the increase in heat when the concrete interacts with the electromagnetic field. Thus, some of the free water is evaporated, leading to the misinterpretation of the amount of remaining water in the concrete.

2.5.4. Drilling/Melting of Concrete

An innovative application of microwave to concrete is the microwave-assisted drilling or melting of concrete. High-energy microwave is applied to an area that is much smaller than the wavelength of the microwave (0.001–1.0 m). The microwave behaves like a "focusing injection," with a monopole antenna conveying the material to be drilled (e.g., concrete, ceramic, etc.; Figure 2.9). The high-energy microwave energy penetrates the small area of the surface of the material, generating a high temperature locally until the material softens or melts. When a near-field radiator is used to induce microwave energy into a small volume on the surface of the material, a hot spot progresses in a rapid thermal runaway process. This process should be taken into account when determining the microwave drilling spot.

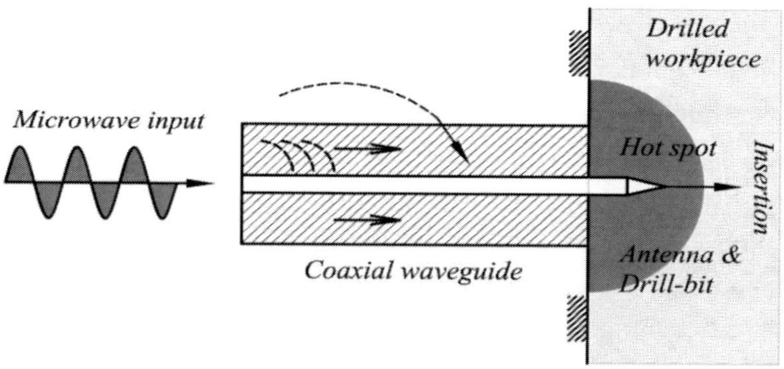

Figure 2.9. A typical microwave drill operating system.

2.6. Theoretical Investigations of Microwave Processing

This section describes an important theoretical investigation point of view of use of microwave in cement and concrete. It consists of electromagnetic theories involved microwave heating by applying Maxwell's equations, electromagnetic power flow by Poything vector, and the local volumetric heat-generation term based on Lambert's law. In addition, numerical techniques and simulation of microwave heating are presented involving partial differential equations (PDEs) of the Maxwell's equations and the heat transport equations.

2.6.1. Electromagnetic Theories

In essence, the numerical simulation/modeling of microwave heating involves the analysis of multiphase coupling. The mathematic model consists of the electromagnetic field equations (i.e., Maxwell's equations) and the heat transport equations, which are interrelated. To capture the interrelationship between the electromagnetic field and heat transport, the heat transport equation contains at least one inhomogeneous term that reflects the heating source provided by the microwave-dissipated power. Temperature variations during heating can cause changes in the complex permittivity, which can lead to space and time variations of the electromagnetic field. To simulate the microwave heating process, the electromagnetic and heat transport equations with their coupled relationships must be modeled together and solved by suitable numerical methods.

The microwave field includes electric ($\vec{E}$) and magnetic ($\vec{H}$) fields that move simultaneously through space. The relationship between these time-varying fields can be described macroscopically with Maxwell's equations, revealing the following important observations: (a) the displacement current density is the rate at which the electric flux density ($\vec{D}$) varies with time; (b) time variations in the electric field create time variations in the magnetic field, due to the action of ($\partial\vec{D}/\partial t$) as a source of the magnetic field; and (c) the time-varying magnetic and electric fields are interdependent. With the help of boundary characteristics, the propagation conditions of plane waves can be assumed as follows: (a) the components of the field quantities ($\vec{E}$ and $\vec{H}$) lie in a transverse plane that is perpendicular to the direction of wave propagation; and (b) the attenuation of the wave amplitude is reduced by a factor of e^{-1} (0.368) in a dielectric medium (i.e., a material that can absorb microwave energy and then can be converted partially into other form of energy), by creating ionic oscillations and continuing dipole relaxation.

The electromagnetic (microwave) waves propagate in free space at the speed of light. Further information on the form of these waves can be obtained by noting that the electric and magnetic fields must also satisfy Maxwell's equations. Although the results derived are general, a specific type of wave is considered: the wave is an unbounded plane, indicating that there exist planes (or wave fronts) that are perpendicular to the direction of propagation, over which all quantities of the wave are constant.

Importantly, Maxwell's equations can be used to predict electromagnetic waves by considering the compacted forms of four equations. The validity of these equations has been proven by their consistency with all existing experiments of electromagnetic phenomena. The physical meanings of these equations are indicated via their specific differential forms below:

(a) Gauss's law: $\nabla \cdot \vec{D} = q$ (2.5)

(b) Faraday's law: $\nabla \times \vec{E} = -\dfrac{\partial \vec{B}}{\partial t}$ (2.6)

(c) Ampère's law: $\nabla \times \vec{H} = \vec{J} + \dfrac{\partial \vec{D}}{\partial t}$ (2.7)

(d) Maxwell's equations: $\nabla \cdot \vec{B} = 0$ (2.8)

where $\vec{E}$ is the electric field intensity (V/m), $\vec{H}$ is the magnetic field intensity (A/m), $\vec{D}$ is the electric flux density (C/m), $\vec{B}$ is the magnetic flux density (Wb/m^2), t is time, and q is the electric charge density (C/m^3).

Constitutive relations that relate $\vec{H}$ and $\vec{B}$, $\vec{E}$ and $\vec{D}$, and $\vec{E}$ and $\vec{J}$ are shown in Eqs. (2.9) to (2.11), respectively. Substituting Eqs. (2.9) to (2.11) into Eqs. (2.5) to (2.8) results in Eqs. (2.12) to (2.15):

$$\vec{B} = \mu\vec{H} \tag{2.9}$$

$$\vec{D} = \varepsilon\vec{E} \tag{2.10}$$

$$\vec{J} = \sigma\vec{E} \tag{2.11}$$

where μ is the magnetic permeability (H/m), σ is the electric conductivity (1/Ohm), and ε is the dielectric constant (F/m).

$$\nabla \cdot \vec{E} = \frac{q}{\varepsilon} \tag{2.12}$$

$$\nabla \times \vec{E} = -\mu\frac{\partial \vec{H}}{\partial t} \tag{2.13}$$

$$\nabla \times \vec{H} = \sigma\vec{E} + \varepsilon\frac{\partial \vec{E}}{\partial t} \tag{2.14}$$

$$\nabla \cdot \vec{H} = 0 \tag{2.15}$$

During microwave heating, electromagnetic energy is converted into heat, which is liberated into or from the dielectric material. Therefore, the power that flows through a closed medium surface can be calculated from the Poything vector, as shown in integral form in Eq. (2.16).

$$\vec{P} = \vec{E} \times \vec{H} \tag{2.16}$$

However, by definition, the average power is given by Metaxas and Meredith:

$$P_{av} = -\frac{1}{2}\int_{s} \mathrm{Re}\,al(\vec{E} \times \vec{H}) \cdot d\vec{S}' = -\frac{1}{2}\omega\varepsilon_0\varepsilon'' \int_{V} (\vec{E}^* \cdot \vec{E})dV \tag{2.17}$$

The temperature change of the processed cement paste during microwave treatment can be obtained by solving the heat-conduction transport equation, in which the microwave power is included as a local electromagnetic heat-generation term:

$$\frac{\partial T}{\partial t} = a\left(\frac{\partial^2 T}{\partial x^2} + \frac{\partial^2 T}{\partial y^2} + \frac{\partial^2 T}{\partial z^2}\right) + \frac{Q}{\rho \cdot C_p} \tag{2.18}$$

In this equation, T is the temperature (°C), a is the thermal diffusivity (m^2/s), ρ is the density (kg/m^3), and c_p is the heat capacity at constant pressure (J/(kg·K)). The local electromagnetic heat-generation term Q depends on the electric field distribution and based on Lambert's law, the microwave energy absorbed or the local volumetric heat-generation term, defined as in Eq. (2.19).

$$Q = \sigma\left|\vec{E}\right|^2 = 2\pi f \varepsilon_0 \varepsilon_r' \left(\tan\delta\right)\left|\vec{E}\right|^2 \tag{2.19}$$

where Q is the microwave energy, σ is the effective conductivity, f is the frequency (Hz), ε_0 is the permittivity of free space (8.8514×10^{-12} Farad/meter), ε_r' is the relative dielectric constant, $tan\ \delta$ is the loss tangent coefficient, and $\vec{E}$ is the electric field intensity (Volt/meter).

When microwave energy penetrates a dielectric material, the wave strength fades away exponentially because the microwave energy is absorbed by the dielectric material and is converted to heat. This absorbed ratio varies with frequency and dielectric properties. In general, the penetration depth (D_p) is used to denote the depth at the power density. It decreased to 36.8% or (e^{-1}) of its initial value at the surface and is defined in Eq. (2.20):

$$D_p = \frac{1}{\frac{2\pi f}{\upsilon}\sqrt{\frac{\varepsilon_r'\left(\sqrt{\left(1+\left(\frac{\varepsilon_r''}{\varepsilon_r'}\right)^2\right)}-1\right)}{2}}} = \frac{1}{\frac{2\pi f}{\upsilon}\sqrt{\frac{\varepsilon_r'\left(\sqrt{\left(1+(tan\delta)^2\right)}-1\right)}{2}}} \tag{2.20}$$

where D_p is the penetration depth, ε_r'' is the relative dielectric loss factor, and υ is the microwave speed in the dielectric material that can be evaluated by $c/\sqrt{\varepsilon_r'}$.

2.7. Numerical Techniques and Simulation

The numerical simulation of microwave heating involves an analysis of at least two kinds of interrelated partial differential equations (PDEs): the electromagnetic field equations (i.e., Maxwell's equations) and the heat transport equations. Additional differential equations may also be needed; for example, the mass transfer equation may be required for a model of drying porous materials.

To obtain an accurate solution for the above coupled equations, analytical methods cannot be used. Various numerical methods have been applied to simulate microwave heating of cement and concrete. Popular methods include Finite Difference (FD) methods (e.g., Finite Differential Time Domain/FDTD method), Finite Element Method (FEM), the Moment of Method (MoM), Transmission Line Matrix (TLM) method, Finite-Volume Time-Domain (FVTD) method, and others. Remarkable progress has been achieved in recent years in the numerical simulation of microwave heating. As more refined models, faster numerical methods, and more complete descriptions of the complex permittivity of more materials are developed, this progress will continue to accelerate. With the aid of numerical simulation, the complex interaction between microwaves and materials will be better understood, and the critical parameters influencing the process will be identified, allowing for the rapid design of optimized and controllable applicators.

To study the microwave heating process in cement and concrete, this process can be separated into a heat transfer model and a microwave power dissipation model, as shown schematically in Figure 10(a). FDTD is a powerful numerical method for solving Maxwell's equations. The FDTD method described by Rattanadecho et al. is a useful approach for solving models in the single-mode cavity, as shown schematically in Figure 2.10(b). Natt et al. used the FEM method to solve Maxwell's equations and heat transfer in the cement paste during microwave curing in the multimode cavity (Figure 2.11). However, Lambert's law cannot be used in numerous situations, such as instances with standing waves.

2.8. Health and Safety of Microwave Processing

Health and safety of microwave processing cannot be neglected when microwave operates with human and environments. Therefore this issue is explained in this section that entails on the limitations and conditions of microwave exposure regulated by associations relating to the use of microwave energy.

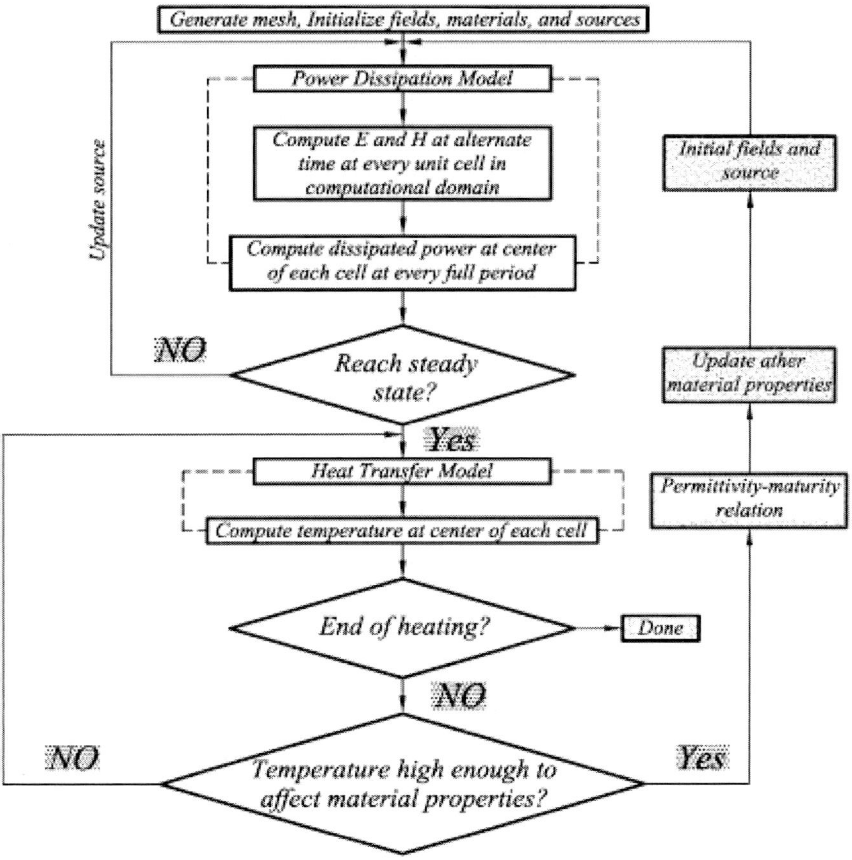

(a) Multi-mode cavity

Figure 2.10. (Continued).

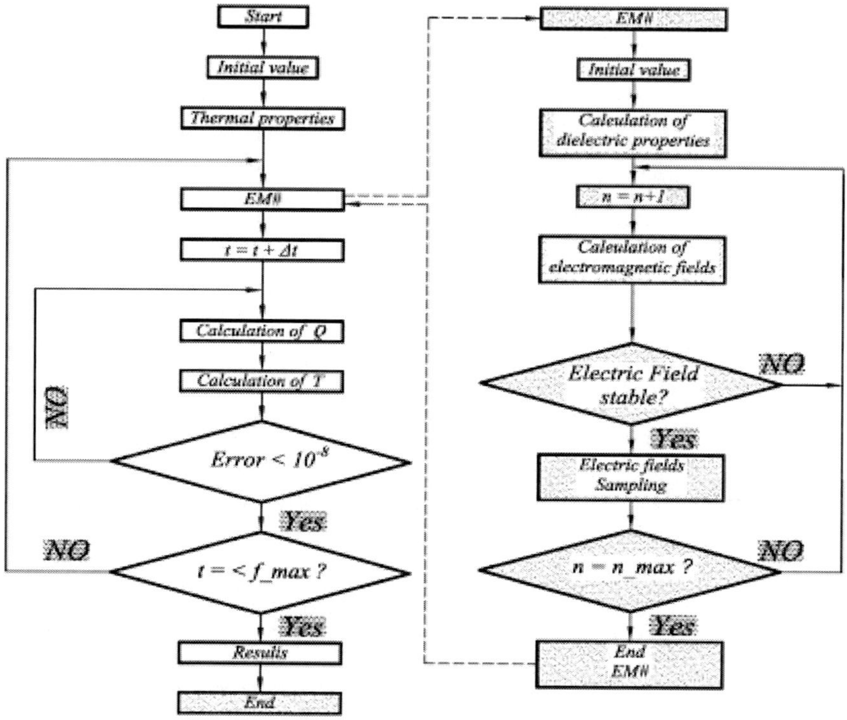

(b) Single-mode cavity

Figure 2.10. Microwave heating model. (a) Multi-mode cavity (b) Single mode cavity.

Microwave processing of cement and concrete materials may involve a risk of exposure of microwave to human and biological tissues if certain precautions are not taken. Therefore, the Federal Communication Commission (FCC) and Occupational Health and Safety Administration (OSHA) have provided exposure limits for microwave. These organizations have considered the effects of microwave radiation on human health from the perspectives of exposure source (e.g., medical devices, amateur radio, cellular phone base stations, hand-held cell phones, heating and sealing devices, microwave ovens, radio broadcast antennae, traffic radar devices, etc.), hazard location, exposure time and solution.

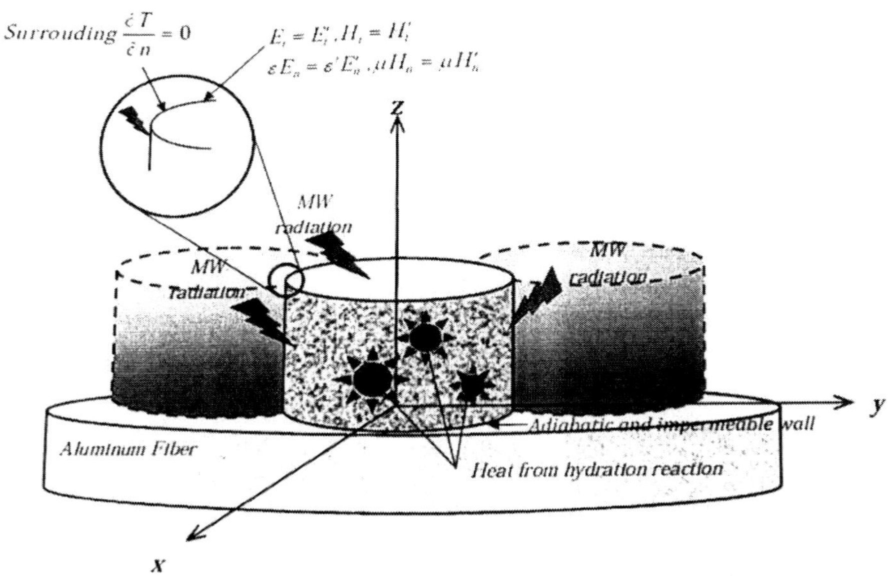

Figure 2.11. A physical model and boundary conditions of 0.38-w/c cement paste subjected to microwave energy using a multi-mode cavity.

The absorption and distribution of microwave exposure are strongly dependent on body size and orientation, as well as on the frequency and polarization of the incident radiation. In particular, the specific absorption rate (SAR) refers to the power absorbed by a body when exposed to EM radiation. The current SAR limit is 4 W/kg. To be conservative, organizations generally use 1/10 of this value for setting standards, especially for radio frequency (RF) and microwave exposure limits. For occupational exposure, the field strength is averaged over a 6-min time period. For public exposure, the field strength is averaged over a 30-min period, as recommended by the International Radiation Protection Association (IRPA) Guidelines, IEEE Standards, National Radiological Protection Board (NRPB), and the Health Protection Agency. It is strongly recommended that the microwave systems in the laboratory and industry are periodically tested for any leakage, and if any substantial leakage is observed, suitable measures must be taken to stop the leakage before the system is further used.

2.9. Conclusions

2.9.1. Major Review Findings

In this review paper, we have described applications of microwave energy in the cement and concrete field. Microwave heating is a high-performance volumetric heating method suitable for dielectric materials used in various concrete industrial processes. Applications of microwave heating in these systems include cement synthesis, accelerated curing, decommissioning of decontaminated concrete, nondestructive monitoring, and concrete drilling/melting. Numerous studies have experimentally and numerically investigated these processes, to utilize microwave heating to its fullest potential in cement and concrete applications. The following conclusions can be drawn from this review:

- Microwave heating is commonly performed at 2.45 GHz. Concrete mixtures can be monitored during the hydration process with a custom-designed parallel-plate setup using a network analyzer.
- The dielectric permittivity of cement-based materials is affected by the initial w/c ratio, cement type, mineral admixture, and aggregate. Although the volumetric fractions of water and superplasticizer in a given mixture are small, these components strongly affect the dielectric permittivity.
- The dielectric permittivity is relatively high and remains constant during the dormant period of each process. The dielectric permittivity decreases rapidly when the hydration reaction resumes, and it continues to decrease during the acceleratory period. Thus, the dielectric property is related to the temperature rise and setting time behaviors of cement-based materials. This information can be used to accelerate the early

strength of the concrete by choosing a suitable curing time (e.g., start time and end time for applying microwave energy).

- The clinkering temperatures required by microwave processing were about 100°C less than those required for fast-heating by the electric furnace method.
- Microwave-assisted heating for 30 to 60 min induced C_3S formation. Ion oxides played a very important role in C_3S formation in conventional sintering.
- Microwave-assisted sintering improved the formation of C_3S compared to conventional sintering methods
- Microwave can be utilized to improve the cementitious properties of RHA-incorporating cement, especially cement that is microwave-sintered at 800°C.
- Use of microwave energy accelerated strength development in Portland cement mortar without decreasing the long-term strength of the concrete. The strength of the concrete specimens began to decrease when the final w/c ratio was less than 0.40.
- Key parameters for microwave curing include the microwave power, application time, duration of microwave heating, thermal runaway, and overheating within the sample
- Microwave energy at operating frequencies of 0.896, 2.45, 10.6, and 18.0 GHz can be used to decontaminate a concrete surface. At or above 10.6 GHz, the microwave power dissipation was shifted towards the front surface of the concrete.
- Microwave energy can be used for near-field microwave-sensing techniques, such as for determining the natural moisture content, percent absorption, bulk specific gravity of aggregates, hydration kinetics of cement, and water content of fresh concrete.
- High-energy microwave energy (wavelength range: 0.001–1.0 m) is useful for drilling/melting of concrete.

- The numerical simulation of microwave heating involves the analysis of electromagnetic field equations (i.e., Maxwell's equations) and heat transport equations. Simulations should be separated into heat transfer and microwave power dissipation models. FDTD is a powerful numerical method for solving Maxwell's equations.
- The specific absorption rate (SAR) refers to the power absorbed by a body when exposed to electromagnetic radiation. The current SAR limit is 4 W/kg.

2.9.2. Future Trends of Research and Development

Microwave heating is a high-performance volumetric heating method suitable for the dielectric materials used in various concrete industrial processes. Numerous studies have experimentally and numerically investigated this process, in order to utilize microwave heating to its fullest potential in cement and concrete applications. Some of the investigated topics are outlined as follows:

- The dielectric properties of cementitious materials should clarify the effects of the time-dependent dielectric properties and thermal properties of concrete on microwave heating mechanisms. The dielectric property is a primary factor for determining the interaction of the microwave energy with the material. Theoretically, the complex permittivity comprises the normal dielectric constant ε', which is a measure of how much energy from an external electric field is stored in a material, and the loss factor ε''·, which is a measure of how much energy a material loses to an external electric field. In addition, concrete is a heterogeneous material with pores that are partially filled by Si^{3+}, Ca^{2+}, and OH^- ion-rich solution. The ion concentration parameters can vary throughout the hydration and pozzolanic reactions. However, heating concrete by microwave energy leads to a simultaneous change in the dielectric property, which instantaneously affects the interaction between the microwave energy and concrete. Thus, the dielectric property needs to be understood, in order to determine the compatibility of microwave applications with the material's time-dependent concrete properties and thermal properties.

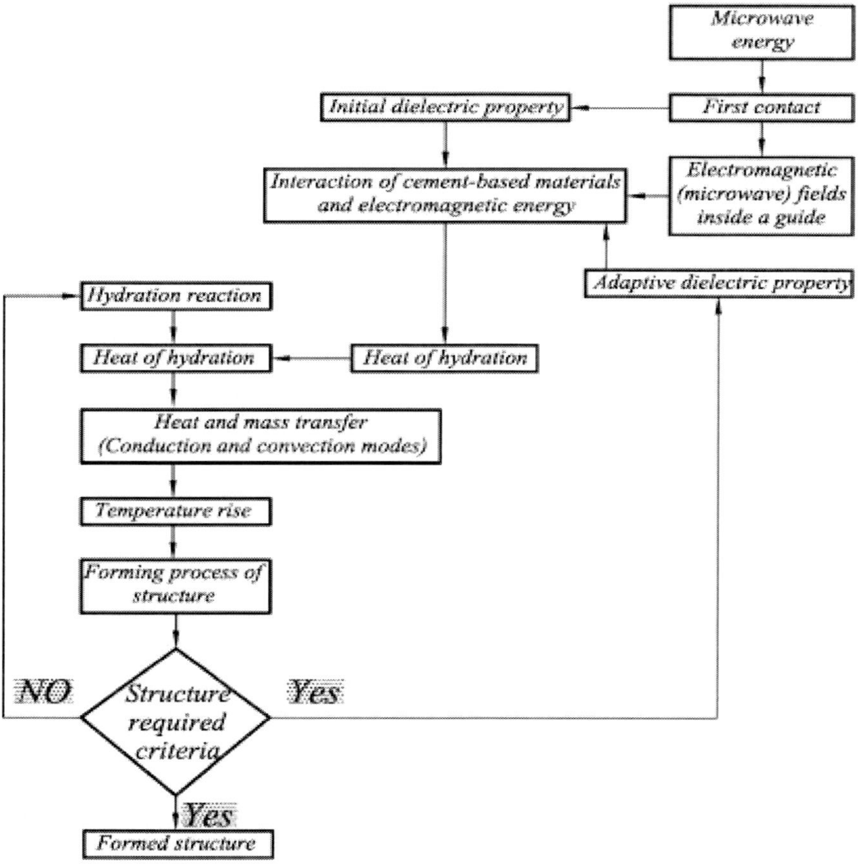

Figure 2.12. A proposed schematic model of cement and concrete subjected to microwave energy.

- Studies should investigate the influences of different concrete mixtures and curing conditions on the characteristics of microwave heating, as these conditions strongly affect the strength development of concrete. The primary parameters that characterize a given concrete sample include the w/c ratio, the chemical composition of the cement, the "fineness" of the cement (e.g., type 1 – normal purpose; type 2 – moderate heat- and sulphate-resistant; type 3 – high early strength; type 4 – low heat; and type 5 – sulphate-resistant), the nature and proportions of aggregates, and the use of chemical and mineral admixtures, among others. Admixtures are increasingly being used in the concrete construction industry. For example, type 1 cement can be replaced, in part, by silica fume, producing a concrete with high early strength.
- As shown in Figure 2.12, Heat and mass transfer mechanisms within the heated concrete during microwave heating should be clarified. Studies have consistently shown that the generation and transport of heat and mass within the heated concrete material are key factors influencing the efficiency of microwave heating. In this process, heat is generated inside the concrete structure, and volumetrically transported towards the outside. These studies assume that no chemical reactions occur within the concrete under consideration. This assumption is not true, however, as hydration and pozzolanic reactions take place. Given the difficulty of this task, one may choose to perform the analysis at specified coupled chemical reactions during microwave curing. Heat generated by the microwave ($Q=\sigma\left|\vec{E}\right|^2 = 2\pi f\varepsilon_0\varepsilon_r'\left(\tan\delta\right)\left|\vec{E}\right|^2$) must be applied in synergy with the heat liberated by the hydration (Q_{Hyd}^{Total}) and pozzolanic (Q_{Pozz}^{Total}) reactions ($Q_{Reaction}^{Total} = Q_{Hyd}^{Total} + Q_{Pozz}^{Total}$). These heat sources should

be considered from the perspectives of the setting and hardening processes of concrete, which are influenced by the internal energy from hydration and the external energy from microwave energy. Simply speaking, when external energy is added to the concrete after the final setting time, the energy leads to the generation and expansion of microcracks.

Chapter 3

DIELECTRIC PROPERTIES OF CONCRETE

The use of microwave energy to improve the properties of cement-based materials is a relatively new area of research. However, this is a growing area of interest because microwave heating has many advantages, including high speed for heat generation, high-energy penetration, instantaneous and precise electronic control, and clean process. Due to various conventional curing methods have many limitations. For example, they take a long time to reach the strength required for water curing. In addition, due to the inherent thermal insulation of concrete, they have non-uniform hydration products; this causes different temperatures to occur in processed concrete under high-stream and temperature-curing conditions. Therefore, numerous research studies have clearly indicated that curing by microwave can further increase the early-age strength of cementitious materials. However, some aspects have not been taken into account. Thus, this chapter presents an investigation the dielectric permittivity of cement-based materials during the initial period of hydration at a frequency of 2.45 GHz.

3.1. Fundamental of the Dielectric Properties of Microwave and Cement-Based Materials

The term microwave heating is equally applicable to microwave systems in both cases; for both, the heating is due to the fact that dielectric material, material with small but finite electrical conductivity, absorbs energy when it is placed in a high-frequency electric field. Consequently, electrical dipole polarization and conduction will be generated within dielectric materials, which are composed of polar molecules with positive and negative poles. These orderly dispersed polar molecules vibrate instantaneously and violently in correspondence to the alternative high-frequency electric field of microwave. It is necessary to overcome the resistance of molecular attraction and motion, because when friction generates heat, the temperature of the material is evaluated simultaneously.

Based on Lambert's law, the microwave energy absorbed can be defined as the root-mean-square value of the electric field intensity $\vec{E}$ is normally used to evaluate the microwave energy absorbed in Eq. (3.1).

$$Q=\sigma\left|\vec{E}\right|^2=2\pi f\varepsilon_0\varepsilon_r^{'}\left(tan\,\delta\right)E^2 \tag{3.1}$$

where Q is the microwave energy, σ is the effective conductivity, f is the frequency, ε_0 is the permittivity of free space (8.8514 x 10^{-12} Farad/meter), $\varepsilon_r^{'}$ is the relative dielectric constant, $tan\ \delta$ is the loss tangent coefficient, and $\vec{E}$ is the electric field intensity.

Since concrete is a dielectric material, its intrinsic properties affect the way it interacts with the electric and magnetic fields of the microwaves. Dielectric property can be characterized by two independent electromagnetic properties: the complex (electric) permittivity ε^* and the complex (magnetic) permeability μ^*. However, most common concrete materials are non-magnetic and yield permeability μ^* that is very close to the permeability of free space ($\mu_o = 4\pi x 10^{-7} Henry/meter$) [16]. Thus, this study focuses on the complex ε^*, which comprises real and imaginary parts as the relationship expressed in Eq. (3.2):

$$\varepsilon_r^* = \varepsilon_r' - \vec{j}\varepsilon_r'' \quad (3.2)$$

where ε_r' and ε_r'' are the real part and imaginary parts respectively, of the complex permittivity, and $\vec{j} = \sqrt{-1}$.

The real part of the relative complex permittivity, referred to as the dielectric constant, ε_r', measures how much energy transferred from an external electric field is stored in a material. The imaginary one, ε_r'', measures how lossy a material is to an external electric field and is referred to as the relative loss factor. Moreover, an essential ratio that shows the energy lost (relative loss factor) to the energy stored (relative dielectric constant) in a material is given as loss tangent $tan\, \delta$, as shown in Eq. (3.3):

$$tan\, \delta = \varepsilon_r'' / \varepsilon_r' \quad (3.3)$$

As discussed earlier the dielectric property plays an important role to describe the behavior of concrete materials when subjected to microwave energy. Research works have been investigated continuously on basic dielectric properties. However, some aspects have not been taken into account. Thus, this paper investigates the dielectric permittivity of cement-based materials during the initial period of hydration at a frequency of 2.45 GHz. The investigation uses a network analyzer with a transmission line technique based on the influences of w/c ratios, cement types, pozzolan materials, and aggregates.

3.2. Materials and Methods

3.2.1. Materials Used

Types I and III hydraulic Portland cements were used throughout this test. Their chemical composition and physical properties are shown in Table 3.1. The ASTM C 618 classifies the FA as low calcium (Type F). Silica fume (hereinafter referred to as SF) was also used as a high-pozzolanic material in accordance with the ASTM C 1240.

Deionized water with a pH 7.5 and river sand with a fineness modulus of 2.58 and gradation conforming to the ASTM C 33 were mixed in specific proportions. The chemical admixture used superplasticizer that conforms to the ASTM C 494; that is, the superplasticizer had a recommended dosage rate of 1000 ml per 100 of a kilogram of cementitious materials. Mineral admixtures included fly ash (here-in-after referred to as FA) from an electricity power plant that used lignite coal as a raw material to produce a combustion-yielding steam for driving the generator.

Table 3.1. Chemical composition of the materials used (% by mass)

Materials	SiO_2	Al_2O_3	CaO	Fe_2O_3	MgO	K_2O	Na_2O	SO_3	L.O.I
OPC	20.30	5.67	60.43	6.23	3.14	0.90	0.36	2.80	2.80
FA	47.80	22.70	2.18	21.90	0.92	2.15	0.32	0.37	1.44
MK	55.14	43.95	0.08	0.75	0.40	2.10	0.03	0.05	12.50
SF	96.0	0.10	0.12	0.10	0.13	0.45	0.10	-	2.57

Table 3.2. Specific gravity and surface area of the materials used

Materials	Specific Gravity	Specific Surface Area (BET method) (m^2/g)
OPC	3.12	0.85
FA	2.45	0.70
MK	2.25	0.60
SF	2.90	18.02

The sizes of the Portland cement and FA particles as shown in Figure 3.1 were obtained using a micrograph sedigraph instrument. The method is based on X-Ray attenuation of a power, which is suspended in a sample cell and allowed to settle. Studies performed using the Malvern instrument Mastersizer 2000 particle size analyzer have achieved accuracy of ± 1% as designated by the $D_{(v,50)}$ using the Malvern Quality Audit Standard.

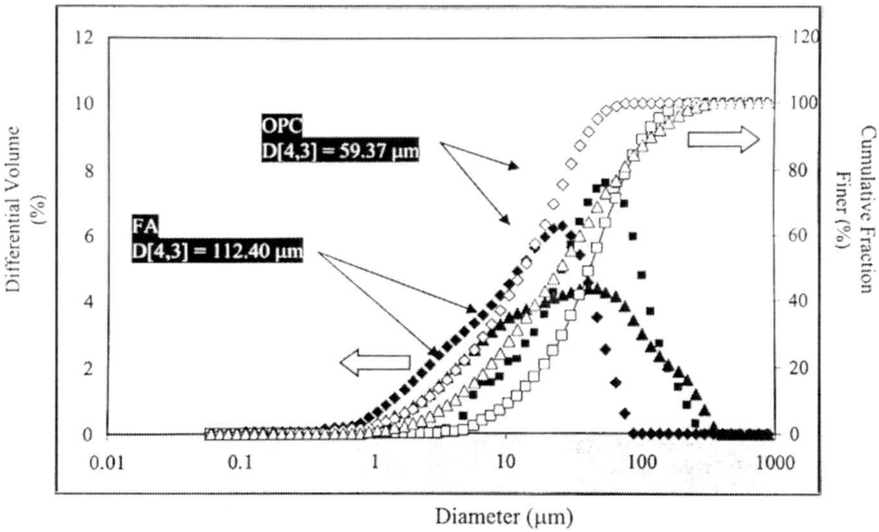

(a) Portland cement Type I OPC and pulverized fuel ash (FA)

Figure 3.1. (Continued).

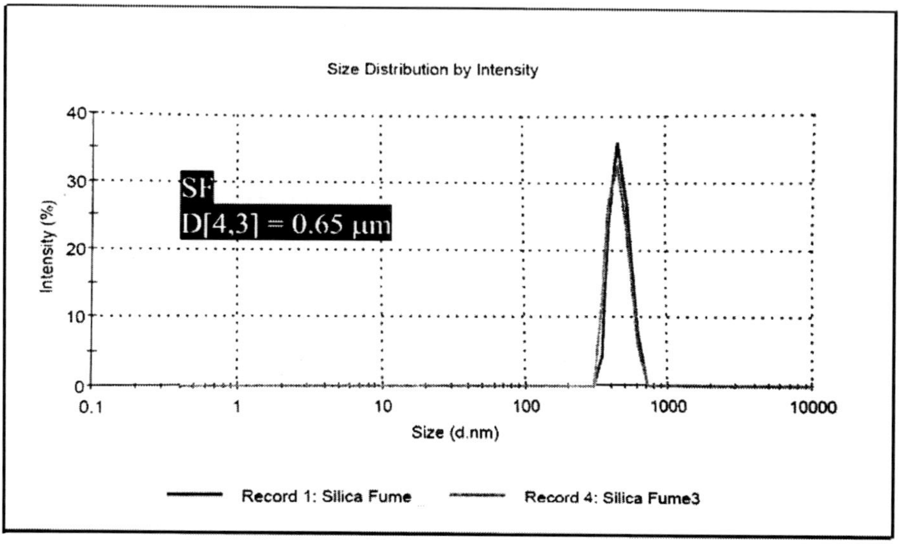

(b) Silica fume (SF)

Figure 3.1. Particle size distributions of Type I Portland cement OPC and FA.

Deionized water with a pH value between 7.0 and 7.5 was used. Ottawa sand, which has a fineness modulus equal to 2.58, constitutes a fine aggregate accurately graded to pass an 850-μ (US Standard No. 20) sieve and to be retained on a 600-μ (US Standard No. 30) sieve—both of which are used to test cements. The primary reason for using Ottawa sand is it's the quality: its grains do not show visible signs of clay, loam, iron compounds, or other foreign substances. Because the sand is typically regarded as a standard, it is often not subjected to detailed mineralogical analyses prior to experiments. However, for this study, we did examine a lab-grade sample in order to identify any impurities present. Crushed limestone rock with a fineness modulus and nominal maximum size equal to 5.74 and 6.0 cm respectively was used as coarse aggregate. Moreover, the test results of specific gravity and absorption of aggregates are also listed in Table 3.3.

Table 3.3. Specific gravity and absorption of aggregates

Properties	Crushed limestone rock	River sand
Bulk Specific Gravity (SSD basis)	2.72	2.34
Bulk Specific Gravity (Oven-Dried basis)	2.71	2.42
Apparent Specific Gravity (SSD basis)	2.70	2.40
Absorption (%)	0.60	0.81

Chemical admixtures: Well-known commercially as Sikament 520 High Range Water Reducing Admixture, the superplasticizer conforms to the ASTM C 494, as comply with ASTM C 494 Type G. Produced and distributed by the Sika Company, Ltd., this particular superplasticizer does not contain calcium chloride or any other intentionally added chlorides, and it will not initiate or promote the corrosion of steel reinforcement in concrete. For general concrete applications, Sika recommends a dosage rate of between 500 – 1200 ml/100 kg) cementitious materials. If maximum water reduction is required, a dosage up to (1400 ml/100 kg) of cementitious materials can be used. In this case, delayed setting times may occur.

3.2.2. Mixture Proportions

The proportions of cementitious materials mixed in preparation for specimen testing are shown in Table 3.4.

Table 3.4. Mixtures used (by weight in grams)

Group	Factors affecting	Symbol	W/S	OPC	FA	SF	MK	Water	Superplasticizer Type G	Sand	Rock
				(g)	(g)	(g)	(g)	(g)	(cc)	(g)	(g)
1	Water-to-solid mass ratio, w/s	1CW/S_P0.25	0.25	500	0	0	0	125	0	0	0
		1CW/S_P0.38	0.38	500	0	0	0	190	0	0	0
		1CW/S_P0.45	0.45	500	0	0	0	225	0	0	0
2	Pozzolan		0.38	400	100	0	0	190	5	0	0
		2CAP_0.38MK20_Super_10%	0.38	400	0	0	100	190	5	0	0
		2CAP_0.38SF20_Super_10%	0.38	400	0	100	0	190	5	0	0
3	Aggregate, Mortar	3CAM0.38_1:2.75	0.38	500	0	0	0	0	5	1375	0
	Aggregate, Concrete	3CAC0.38_1:1:1	0.38	500	0	0	0	0	5	500	500

3.2.3. Testing Procedures

In order to measure the dielectric properties of the paste, mortar, and concrete, it was necessary to use a vector network analyzer (VNA). To measure the dielectric properties of cementitious materials at a frequency of 2.45 GHz, a network analyzer with an open-ended coaxial probe, as shown in Figure 3.2, was used. The analyzer consisted of a coaxial cavity; microwave reflectometer; 3.5-mm coaxial cable; 3.5-mm female calibration; and short-, open-, matched-load software. The coaxial cavity characterizes measurement in the range of 1.5 - 2.6 GHz with precision not more than 2% of the dielectric constant and 5% of the dielectric loss factor. The measured sample should be assumed; i.e., it should be assumed to have infinite size, non-magnetic material, isotropic and homogeneous properties. In addition, the coaxial cavity must be in contact with the sample under test (MUT). The Nicholson-Ross-Weir conversion process was used to calculated dielectric properties. After the cementitious material had been mixed and placed in the mold, it was wrapped in Styrofoam that was 5.0 mm thick in order to protect it from heat loss. Both dielectric properties and semi-adiabatic temperature using a data logger with thermo-couple (Type K) was simultaneously recorded every 180 and 15 min, respectively. However, in order to eliminate the effect of the thermo-couple embedded in microwave radiation, three samples were separately tested for dielectric properties and three for temperature rise.

A measurement using the open ended probe technique involves measuring the reflected signal (S_{11}) and the transmitted signal (S_{21}). The relevant scattering parameters relate closely to the complex permittivity and permeability of the material by equations. The conversion of s-parameters to complex dielectric parameters is computed by solving the equations using a Nicholson-Ross-Weir (NRW) technique.

In order to obtain the permittivity and permeability from s-parameters, the Nicholson–Ross–Weir (NRW) technique (Figure 3.3) was used to provide a direct calculation of both the permittivity and permeability from the s-parameters. The NRW technique is divergent at integral multiples one-half wavelength in the sample. This is due to the fact that at a point corresponding to the one-half wavelength, the s-parameter (S_{11}) gets very small. For a small S-parameter (S_{11}) value the uncertainty in the measurement of the phase of S_{11} on the VNA is very large. Therefore, the uncertainty causes a divergence at these frequencies. Such divergences can be avoided by reducing the sample length, but it is difficult to determine the appropriate sample length when the sample's ε and μ are unknown. Measurement of the reflection coefficient and transmission coefficient requires a pair (S_{11}, S_{21}) of s-parameters of the material to be measured. However, due to phase ambiguity, the technique diverges for low-loss materials at frequencies corresponding to integer multiples of one-half wavelength in the sample. Hence, it is restricted to optimum sample thickness of $\lambda_g/4$ and used preferably for short samples.

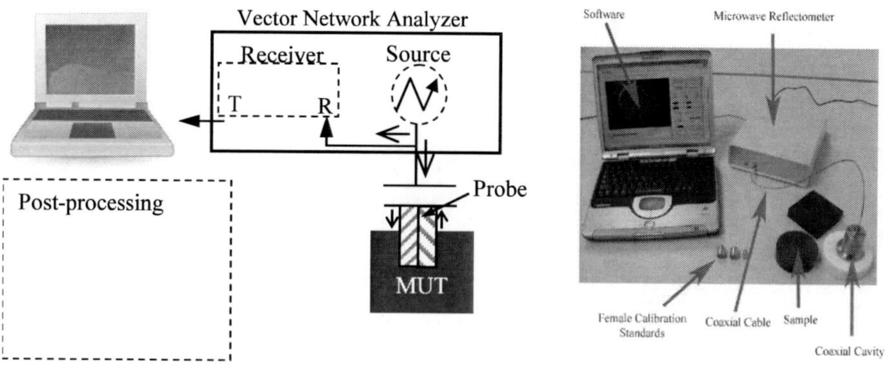

Figure 3.2. A network analyzer (open ended probe technique) (R = reflected power, T = transmitted power).

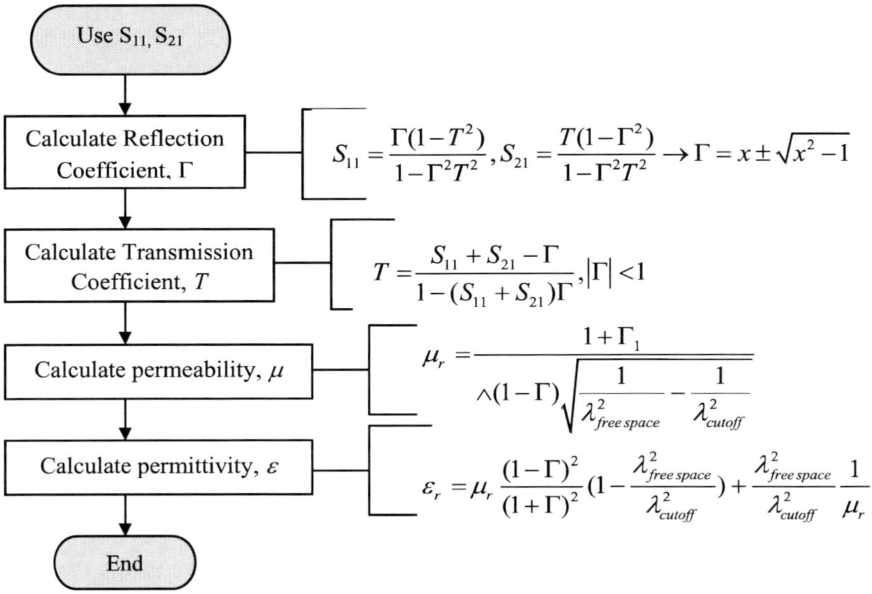

Figure 3.3. Nicholson–Ross–Weir for calculating dielectric permittivity.

3.3. Results and Discussion

All measured waveforms were obtained in the time domain and a frequency range from 1.7-2.6 GHz (S-band). They were then converted into the dielectric permittivity (dielectric constant, ε' and loss factor, ε'') were calculated. The effect of water-cement ratio (w/c), pozzolan materials, aggregate type (sand with and without crushed limestone rock on its properties was studied and also discussed in this section.

3.3.1. Scattering Parameters

Theoretically, scattering parameters or S-parameters describes the electrical behavior of linear electrical networks when undergoing various steady state stimuli by small signals. They are mostly used for networks operating at radio frequency (*RF*) and *microwave* frequencies where signal power and energy considerations are more easily quantified than currents and voltages. Conventionally, s-parameters are defined for a given frequency and system impedance, and vary as a function of frequency as shown typically in Table 3.5, for this study it is assigned in S-band (2.45 GHz). In order to convert from obtained s-parameter from dielectric parameters using a network analyzer, Nicolson-Ross-Weir technique was used due to fast, non-iterative and for applicable to waveguides and coaxial line.

Presence of water is recognized as the primary and dominant factor surrounding the increase of ε' in concrete. Other influencing factors are: electromagnetic (EM) frequency, water-cement ratio, porosity, ions in pore solution which make the material behave as conducting dielectrics, and clay minerals of wide range of porosities and specific surfaces. Factors such as cement type, salt impregnation, presence of fly ash and ground granulated blast furnace slag, and temperature during tests are reported to be of minor or negligible significance. Literature was scarcely reported on the effects of different types of dense aggregate on ε'. However it is believed that the effects should be insignificant because of the relatively small variations of aggregate density.

Table 3.5. IEEE Radar Band Designations

Frequency	Wavelength	IEEE Radar Band designation
1 - 2 GHz	30 - 15 cm	L Band
2 - 4 GHz	15 - 7.5 cm	S Band
4 - 8 GHz	7.5 - 3.75 cm	C Band
8 - 12 GHz	3.75 - 2.50 cm	X Band
12 - 18 GHz	2.5 - 1.67 cm	Ku Band
18 - 27 GHz	1.67 - 1.11 cm	K Band
27 - 40 GHz	11.1 - 7.5 mm	Ka Band
40 - 75 GHz		V Band
75 - 110 GHz		W Band
110 - 300 GHz		mm Band
300 - 3000 GHz		u mm Band

As shown in Figures 3.4 - 3.6, the values of S_{11} and S_{21} values from which were measured in the range of 1.7 – 2.6 GHz of cement pastes at a delay time of 30 min with different water-cement ratios (w/c), with different pozzolan materials and mortar and concrete are represented, respectively. In general, S_{11} and S_{21} represent reflection coefficient at port 1 (inlet signal) when port 2 (outlet signal) in terminated with a matched load and attenuation of wave travelling for port 1 to port 2, respectively.

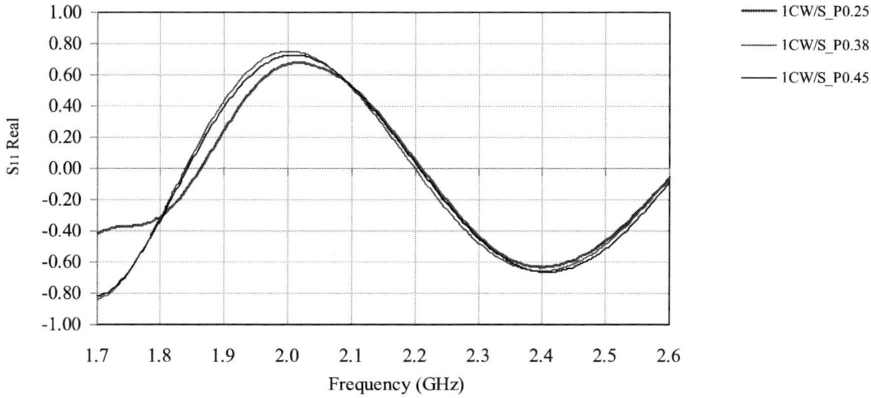

(a) S_{11} real values.

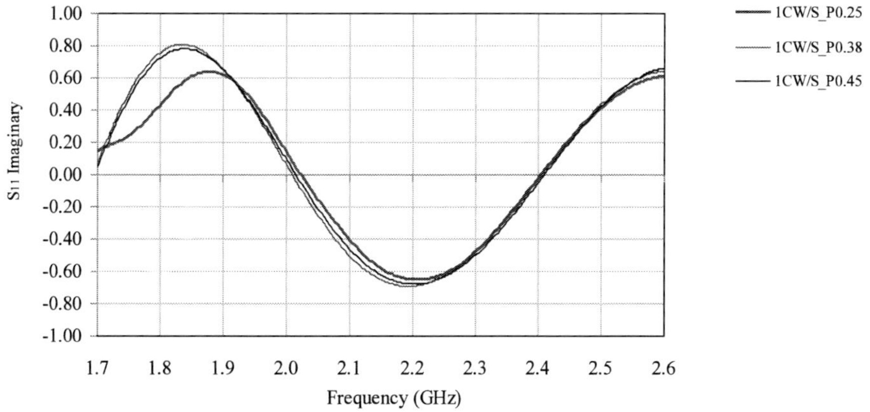

(b) S_{11} imagination values.

Figure 3.4. (Continued).

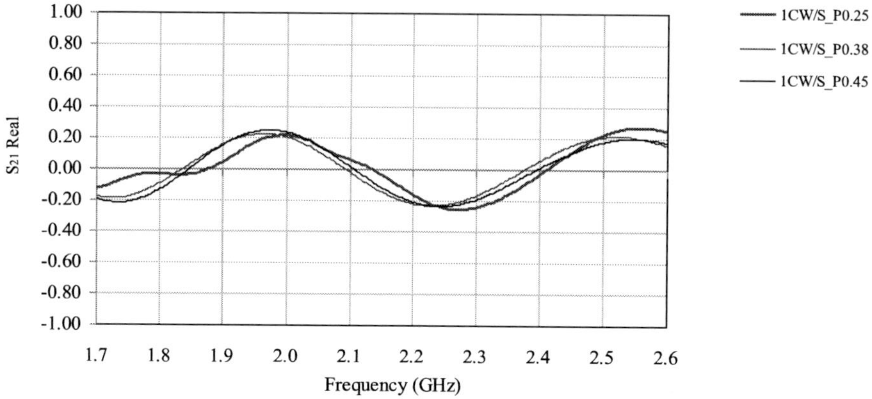

(c) S_{21} real values.

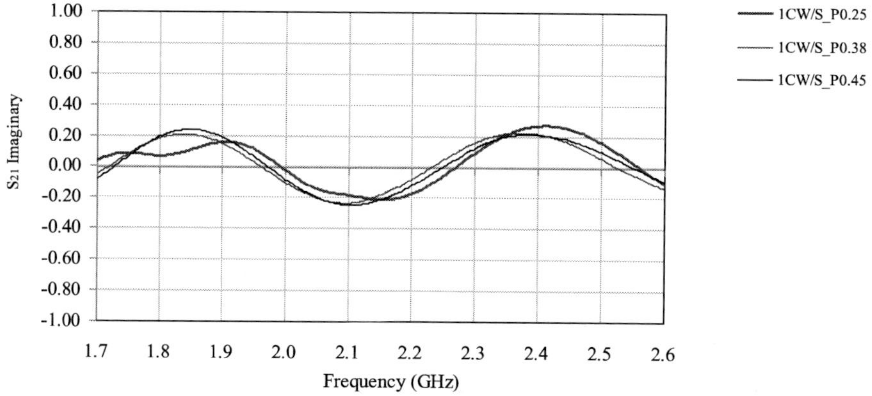

(d) S_{21} imagination values.

Figure 3.4. S_{11} and S_{21} values of cement pastes with different water-to-cement ratios.

For S_{11} and S_{21} values with different water-cement ratios (w/c), the tendency of S_{11} and S_{21} corresponds to the incident wave signal or sinusoidal wave behavior. It can be seen that cement pastes at w/c ratios of 0.38 and 0.45 have similar trend, while the 0.25-w/c cement paste has different value, especially at frequency 1.7 – 1.9 GHz. This may due to the free water content of the paste 0.25 is lower than those of others pastes.

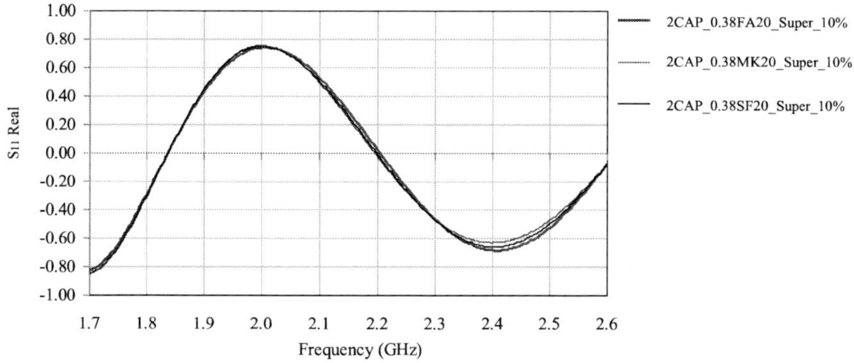

(a) S_{11} real values.

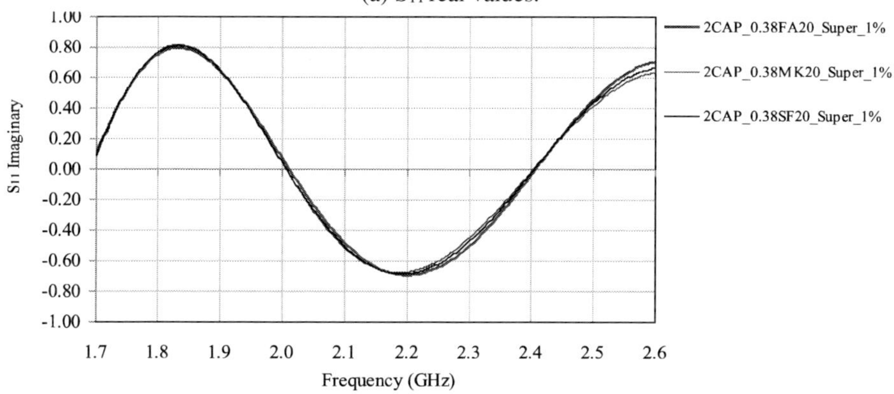

(b) S11 imagination values.

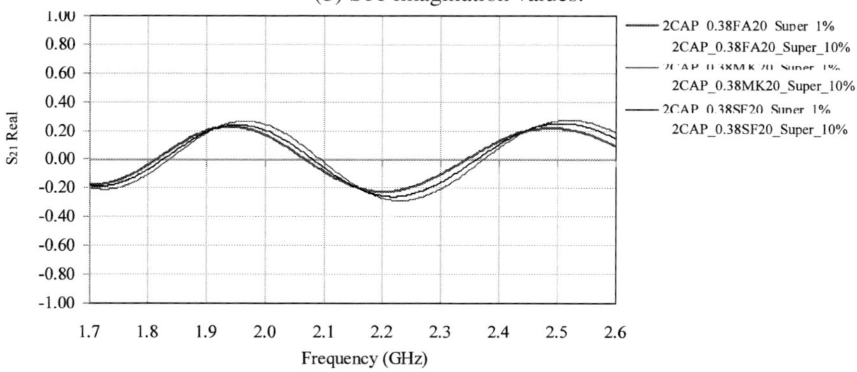

(c) S_{21} real values.

Figure 3.5. (Continued).

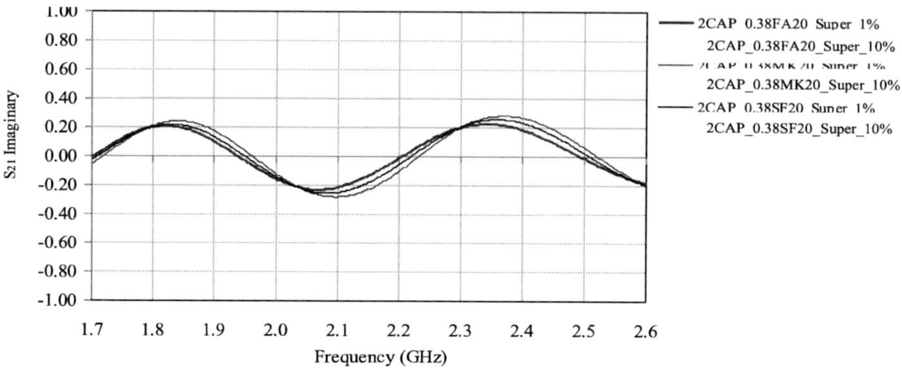

(d) S_{21} imagination values.

Figure 3.5. S_{11} and S_{21} values of cement pastes with different pozzolan materials.

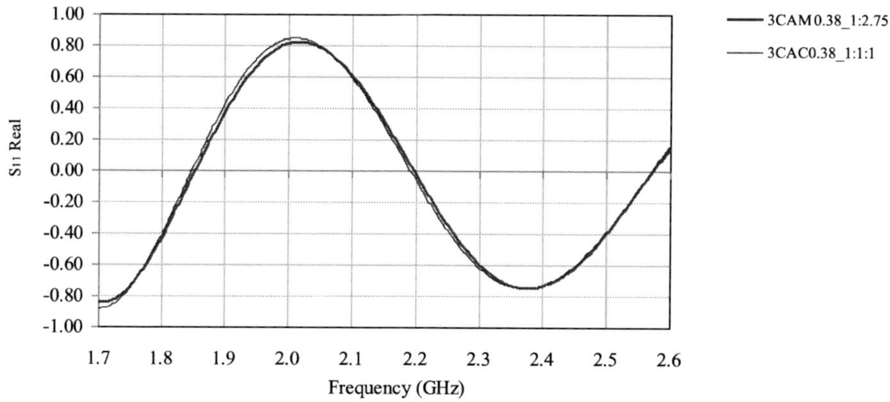

(a) S_{11} real values.

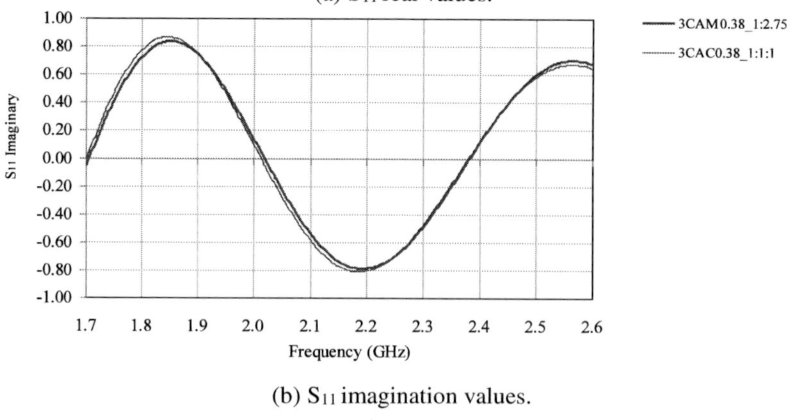

(b) S_{11} imagination values.

Figure 3.6. (Continued).

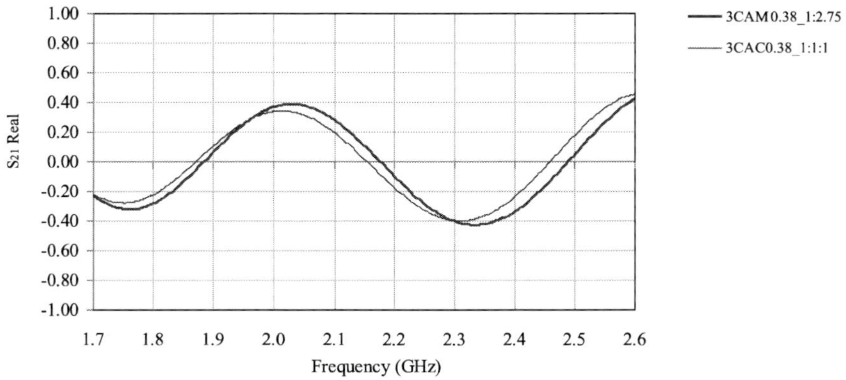

(c) S_{21} real values.

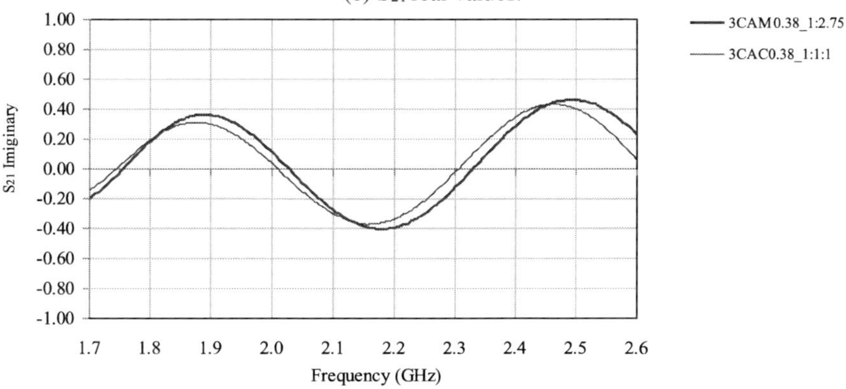

(d) S_{21} imagination values.

Figure 3.6. S_{11} and S_{21} values of mortar and concrete.

3.3.2. Dielectric Permittivity of the Constituents of Cementitious Materials

Table 4.6 shows dielectric permittivity as a function of the temperature of all cement-based constituents. In general, dielectric permittivity consists of dielectric constant (ε') and loss factor (ε'') (Portland cement Types I and III, silica fume, fuel ash, superplasticizer, river sand, and crushed limestone rock) as a function of temperature change. It can be separated into three groups of materials including the following:

(a) The powder material group comprises Portland cement Types I and III, silica fume, and fly ash. This group shows a narrow range of dielectric constant and loss tangent throughout the temperature range of 20 - 50°C. However the loss factor of the materials has a wide variation that is greater that its dielectric constant; this is due to ε' having a high variation and it is related to the conductivity of a material's polarization and relaxation behaviors.
(b) The water-based materials group consists of water and superplasticizer (polycarboxylic water-based). It has a reasonably wide range of both properties. This is because water is a dielectric material with ε' and $\tan\delta$ much higher than that of the other components (Metaxas, 1991). This effect shows clearly in chemical admixtures that consist mainly of water and polymer. In fact, the manufacturer specified that the solid content of superplasticizer is 40% by weight. Thus, the dielectric properties of admixtures are strongly influenced by water content.
(c) The aggregate materials group, for example, the fine aggregate (river sand) and the coarse aggregate (crushed limestone rock), has dielectric properties that are narrower than those of the powder and water-based material groups. For the dielectric constant under the same compaction condition, fine aggregate materials show a lower value than powder materials because the former is looser in its consistency.

Table 3.6. Dielectric constant (ε') and loss factor (ε'') of constituents

Constituents	Dielectric constant (ε'); $\varepsilon'(t)=at^3+bt^2+ct+d$			
	a ± *s.d.*	*b* ± *s.d.*	*c* ± *s.d.*	*d* ± *s.d.*
Portland cement Type I	-3.214E-07 ± 0.251E-06	3.414E-05 ± 0.915E-04	-2.220E-03 ± 0.114E-03	3.215E+00 ± 0.731E-01
Portland cement Type III	-1.033E-07 ± 0.409E-06	3.319E-05 ± 0.881E-04	-1.311E-03 ± 0.245E-03	6.841E+00 ± 1.377E-01
Fly ash	3.614E-07 ± 2.702E-06	-5.231E-05 ± 4.997E-04	4.301E-03 ± 1.224E-02	4.741E+00 ± 0.220E-01
Silica fume	-3.482E-07 ± 6.237E-08	5.614E-05 ± 2.005E-05	-3.140E-03 ± 0.775E-03	2.785E+00 ± 0.678E-01
Superplasticizer (polycarboxylic water-based)	-2.415E-05 ± 7.995E-05	3.995E-03 ± 8.143E-03	-3.881E-01 ± 0.235E-01	2.423E+01 ± 0.874E-01
Fine aggregate (River sand)	-0.952E-08 ± 5.020E-08	-2.155E-06 ± 0.088E-05	2.808E-04 ± 0.214E-04	2.119E+00 ± 0.445E-01
Coarse aggregate (Crushed limestone rock	-7.223E-08 ± 1.414E-07	2.448E-05 ± 0.892E-05	-1.750E-03 ± 0.223E-03	1.346E+00 ± 0.989E-01
Portland cement Type I	2.874E-08 ± 0.902E-08	-7.098E-06 ± 0.098E-05	6.094E-04 ± 0.004E-04	1.064E-01 ± 1.134E-02
Portland cement Type III	7.245E-08 ± 0.885E-08	-3.948E-06 ± 0.083E-06	6.904E-04 ± 2.312E-04	3.033E-01 ± 0.093E-02
Fly ash	-1.775E-07 0.975E-07	2.098E-05 0.048E-05	-1.349E-03 0.904E-03	2.424E-01 7.090E-02
Constituents	Loss factor (ε''); $\varepsilon''(t)=et^3+ft^2+gt+h$			
	e ± *s.d.*	*f* ± *s.d.*	*g* ± *s.d.*	*h* ± *s.d.*
Silica fume	-3.028E-07 ± 0.038E-07	6.438E-05 ± 0.048E-05	-3.948E-03 ± 0.140E-03	2.243E-01 ± 0.3987E-02
Superplasticizer (polycarboxylic water-based)	-2.208E-05 ± 0.849E-05	3.039E-03 ± 0.088E-03	-1.902E-01 ±0.098E-02	9.594E+00 ± 1.759E+00
Fine aggregate (river sand)	7.008E-08 ± 0.084E-08	-2.448E-05 ± 2.949E-06	1.442E-03 ± 0.233E-04	2.244E-01 ± 0.004E-2
Coarse aggregate (crushed limestone rock)	8.044E-09 ± 0.048E-09	-2.330E-06 ± 8.904E-07	2.088E-04 ± 4.234E-05	5.948E-02 ± 0.740E-03

Remark: *s.d.* = Standard Deviation.

3.3.3. Dielectric Permittivity of Pastes, Mortars, and Concretes

Figure 3.7 shows the evolution of dielectric properties and the simultaneous temperature rise of cement pastes. It can be observed that the dielectric properties at the initial stage are relatively higher in comparison with the later stage; they also increase with the increasing water content (higher w/c) of the cement pastes. This is due to the fact that immediately after contact has been made between water and cement, they start to react and then Ca^{2+}, OH^{-}, and SO_4^{2-} ions dissolve into the system. In addition, during the dormant period, the dielectric properties change very little because the chemical composition of the aqueous phase remains nearly constant.

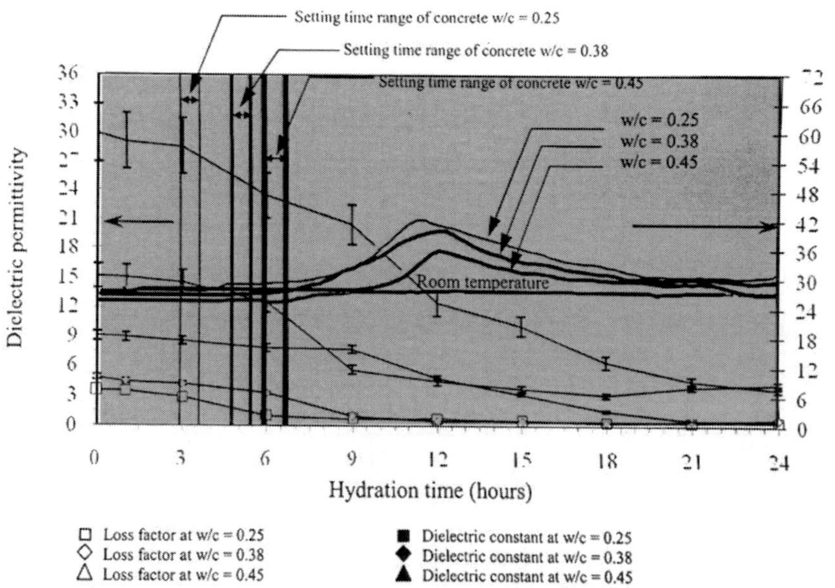

Figure 3.7. Dielectric permittivity of pastes with different water-to-cement ratios.

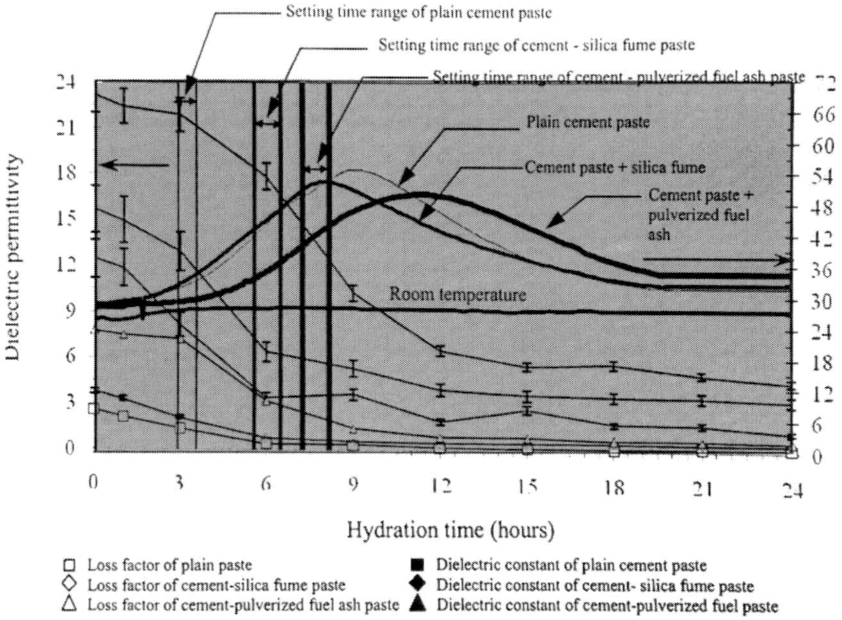

Figure 3.8. Dielectric permittivity of pastes with different pozzolan materials.

Similarly, relative dielectric properties also appear to be affected by the temperature rise. The lower temperature (higher w/c) leads to enhance these properties. This is due to a reduction in water-cement-ratio that accelerates hydration and results in higher temperature. Especially, in the accelerated period of the pastes at which it has the highest rise in temperature corresponding to the decreasing relative dielectric properties.

Effects of silica fume (SF) and fuel ash (FA) on the dielectric permittivity, temperature rise, and setting time of the pastes are shown in Figure 3.8. The dielectric constant of the paste containing FA through the first 24-h hydration time is higher than that of the plain paste, whereas the paste containing SF is lower. Both reactions involving SF and FA occur as secondary reactions. This means, however, that the FA can produce excessive water in the paste while increasing Si^{3+} and Ca^{2+} ions in the system. This results in an increasing dielectric constant of the paste. Unlike the SF paste, the compositions with FA can dissolve it at a high rate; its fineness induces the bounding of its water molecules and ion-richness at the surface. As a result, the dielectric constant is lower than that of the plain cement paste. For loss factor evolution, the conventional paste and the SF paste show little difference in their comparative loss factors; however between, the difference in the loss factors of the conventional and SF paste when compared with the FA paste is large. This indicates that the remaining water content, both during introduction and acceleratory periods in the paste, has a strong effect on dielectric loss. On the other hand, the FA particles can retain free water in the paste system, which may cause the loss factor of the FA paste to become very high.

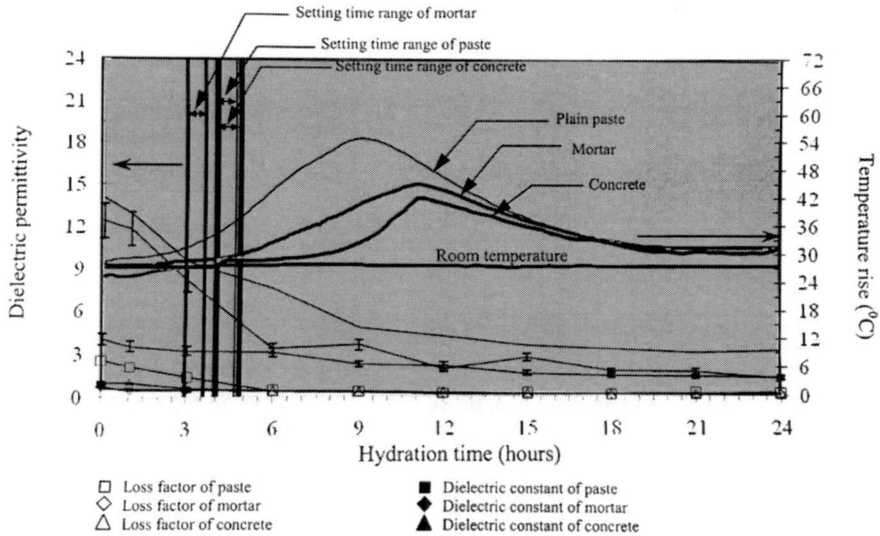

Figure 3.9. Dielectric permittivity of paste, mortar and concrete.

Figure 3.9 shows that the relative permittivity curves for the mortar and cementitious material are also similar to the curve of the conventional pastes but lower than that of the pastes. The decrease of cement content and the absorption of water molecules by aggregate surfaces induce a lower concentration of ions in the system (Rhim and Buyukozturk, 1998). However, eventually, these parameters approach to constant. A comparison in the setting time range shows that the paste's permittivity decrease rate slope is somewhat higher than those of mortars and cementitious materials. This is due to the ion constraints of the hydrated products; specifically that the hydrated products have high ion concentration in comparison with the aggregate-mixed products.

3.4. Conclusions

3.4.1. Dielectric Permittivity

The dielectric permittivity of cement-based materials can be classified into three groups; (a) the powder material group comprises Portland cement Types I and III, silica fume, and fly ash. This group shows a narrow range of dielectric constant and loss tangent throughout the temperature range of 20 - 50°C. (b) The water-based materials group consists of water and superplasticizer (polycarboxylic water-based). It has a reasonably wide range of both properties. (c) The aggregate materials group, the fine aggregate (river sand) and the coarse aggregate, has dielectric properties that are narrower than those of the powder and water-based material groups.

3.4.2. Factors Affecting Dielectric Permittivity

Dielectric permittivity of cement-based material is affected by initial water-cement ratio, cement types, pozzolan and aggregate types. However, although the volumetric fraction of water and superplasticizer in a given mixture is small, it strongly affects the dielectric permittivity of the cement due to its high dielectric permittivity. The change in the dielectric permittivity is relatively high and remains constant during the dormant period; after that it decreases rapidly when the hydration reaction resumes, and it continues to decrease during the acceleratory period.

Chapter 4

MICROWAVE HEATING OF CONCRETES AND REINFORCED CONCRETE

4.1. INTRODUCTION

This chapter presents the microwave-assisted heating of concrete materials and reinforced concrete, i.e., cement pastes containing pozzolan materials, mortars, and concretes are explored in detail here both experimentally and numerically. The first section experimentally investigates the interaction between microwave energy and Type I cement–based materials with and without pozzolan materials during the first 24 h hydration period. The second section investigates the microstructural characteristics of hydration products in cement-based materials subjected to microwave-irradiated energy based on constant microwave-power levels, and compares the results for the same kinds of materials subjected to conventional and high-temperature curing methods. The next chapter will propose a microstructural development for concrete materials' pastes subjected to high-temperature curing using microwave energy.

4.2. Extended Scope of the Research

- The mix proportions of cement-based materials include water-to-solid (Type I Portland cement with/without pozzolan materials; water-cement ratios (w/c) (0.25 (compatible with stoichiometry and consisting of anhydrous cement), 0.38 (fully hydrated paste) and 0.45 (hydration product and capillary pores)). The chemical admixture uses a superplasticizer agent that conforms to the ASTM C 494 Type F (polycarboxylate-based admixture): that is, the superplasticizer agent has a recommended dosage rate of 1000 ml per 100 kg of total solid materials content, which is the replacement level of a pozzolan material; the agent also has fly ash (FA), metakaolinite (MK), and silica fume (SF) as 20%wt. of its solid materials content.
- The surrounding parameters at the water-cement ratio (w/c) of 0.38 include pressure levels at normal pressure (760 ± 5 mm Hg, torr).
- The properties of Type I Portland cement are its composition (X-ray Fluorescence (XRF)) and physical properties (particle size distribution (particle size analyzer)) and surface morphology (Scanning electron microscopy (SEM)).
- Conditions of curing
  - Saturated lime water (temperature: 25 ± 2°C)
  - Microwave energy: constant power levels
    - Delay time of 30 min, initial setting time, and final setting time
    - Power levels: 50 ± 5, 100 ± 5 and 150 ± 5 watt
    - Times of application: 0, 10, 20 and 30 min
- Properties of cement-based materials
  - Macroscopic: strength (axial load compression) at 8 h, 24 h, 7 days and 28 days, heat of hydration and temperature rise

- Microscopic: surface morphology and compositions characteristic of C-S-H, $Ca(OH)_2$ and ettringite (X-ray diffraction), surface morphology and compositions (SEM-EDX), and thermogravimetric analysis (TGA-DTA)

- Microwave-heating model of cement paste: Heat-generation (conduction mode) model.

4.3. MATERIALS AND METHODS

4.3.1. Materials Used

The starting materials used in this study included several commercial cementitious materials, i.e., Portland cement Type I, fly ash (FA), metakaolin (MK), and silica fume (SF). A variety of characterization techniques was used to determine the chemical composition and physical properties of these materials. Roll bar 12.0 mm diameter was used as reinforced steel embedded in cementitious materials.

4.3.1.1. Characterization of Starting Cementitious Materials

The chemical composition of the Portland cement Type I OPC, FA, (CLASS F), MK, and condensed SF as weight percent was determined by means of a DC argon plasma spectrometer and trimetric analysis.

Deionized water with a pH value of 7.5 was used. Ottawa sand, which has a fineness modulus equal to 2.58, constitutes a fine aggregate accurately graded to pass an 850-μ (US Standard No. 20) sieve and to be retained on a 600-μ (US Standard No. 30) sieve—both of which are used to test cements. Crushed limestone rock with a fineness modulus and nominal maximum size equal to 5.75 and 5.90 cm respectively was used as coarse aggregate. Moreover, the test results of specific gravity and absorption of aggregates are also listed in Table 3.2.

Chemical admixtures: Well-known commercially as Sikament 520 High Range Water Reducing Admixture, the superplasticizer conforms to the ASTM C 494, as comply with ASTM C 494 Type G. For general concrete applications, Sika recommends a dosage rate of between 500 – 1200 ml/100 kg) cementitious materials. If maximum water reduction is required, a dosage of 1400 ml/100 kg) of cementitious materials can be used. In this case, delayed setting times may occur.

4.3.2. Mixture Proportions

The mixture proportioning was based on the above concepts. Table 4.1 shows the pastes, mortar, and concrete mixture. In addition, the water-cement mass ratio (w/c), pozzolan, aggregate mortar and concrete were studied. Table 3.4 show the mixtures classified according to the curing method by which they are traditionally cured, i.e., conventional curing and rapid curing using microwave energy. In particular, the group of mixtures that are typically cured using microwave energy, the water-cement ratio of 0.38 by weight was emphasized because it is well-known that this value consists entirely of hydration products.

4.3.3. Mixing and Molding

The appropriate amounts of starting materials (Table 4.1) were weighed out to the nearest hundredth of a gram on a Mettler Pl 1200 balance. A Hobart mixer was used to mix the solids and liquids according to ASTM C 109 (ASTM, 2009). Samples were cast 109.22 mm long, 54.61 mm wide, and 50 mm thick specimens for paste, mortar, and concrete. The samples were cured by using saturated lime water at 25, microwave energy with a multimode cavity, or autoclaving.

The setting time was measured in accordance with ASTM C 191. Pastes were cast normally, and their compressive strengths were determined using 109.20 mm long, 54.60 mm wide, and 50 mm thick cube specimens. As well as, cement paste specimens, mortar specimens: Ottawa sand ratio was 1: 2.75. The water-to-solid ratio was 0.38. For concrete, the cementitious materials: Ottawa sand: crushed limestone rock was 1:1:1. The water-cement ratio was 0.38. The specimens were removed from the mold 23½ ± ½ h after the start of mixing and then cured by conventional curing, rapid curing with microwave energy, and rapid curing: autoclave curing. The specimens subjected to conventional curing were immersed in a $Ca(OH)_2$ saturated solution.

Table 4.1. Mixtures used (by weight in grams) with microwave energy

Group	Factors affecting	Symbol	W/S	OPC	FA	SF	MK	Water	Superplasticizer Type G	Sand	Rock
				(g)	(g)	(g)	(g)	(g)	(g)	(g)	(g)
1	Dormant period	1DP_50(30 min, initial set, final set)	0.38	500	0	0	0	190	0	0	0
		1DP_100(30 min, initial set, final set)	0.38	500	0	0	0	190	0	0	0
		1DP_150(30 min, initial set, final set)	0.38	500	0	0	0	190	0	0	0
2	Accelerated period	2AP_50 (10,20,30)	0.38	500	0	0	0	190	0	0	0
		2AP_100(10,20,30)	0.38	500	0	0	0	190	0	0	0
		2AP_150(10,20,30)	0.38	500	0	0	0	190	0	0	0
3	Sequential, Power level (time of application)	50(10)100(10)50(10)	0.38	500	0	0	0	190	0	0	0
		150(10)100(10)50(10)	0.38	500	0	0	0	190	0	0	0
		100(20)50(10)	0.38	500	0	0	0	190	0	0	0
		50(10)100(10)150(10)	0.38	500	0	0	0	190	0	0	0
		100(10)150(10)50(10)	0.38	500	0	0	0	190	0	0	0
		100(10)150(10)100(10)	0.38	500	0	0	0	190	0	0	0
		50(10)150(10)50(10)	0.38	500	0	0	0	190	0	0	0

4.3.4. Microwave Curing Setup

For accelerated curing, the microwave system used was a monochromatic microwave at a frequency of 2.45 GHz, as shown in Figure 4.1. Microwave energy was generated by a magnetron and transmitted directly along the propagation direction (+z) of a rectangular wave guide toward a water load situated at the end of the waveguide to ensure that a minimal amount of microwave energy would be reflected back to the sample. A warming water load was circulated through the cooling tower in order to reduce the temperature in the water load system.

A cement paste sample was arranged perpendicular to the propagation direction. A Type K thermocouple with a 0.1 mm diameter was inserted at the center of the sample for the purpose of monitoring the temperature rise. During a 10-min period of microwave heating, the output of the microwave magnetron was controlled at 50, 100 and 150 W. The microwave plane wave traveled directly along the wave guide and made contact with the sample surface; the wave was then reflected and transmitted. By using a wattmeter, incident, reflected and transmitted waves were monitored.

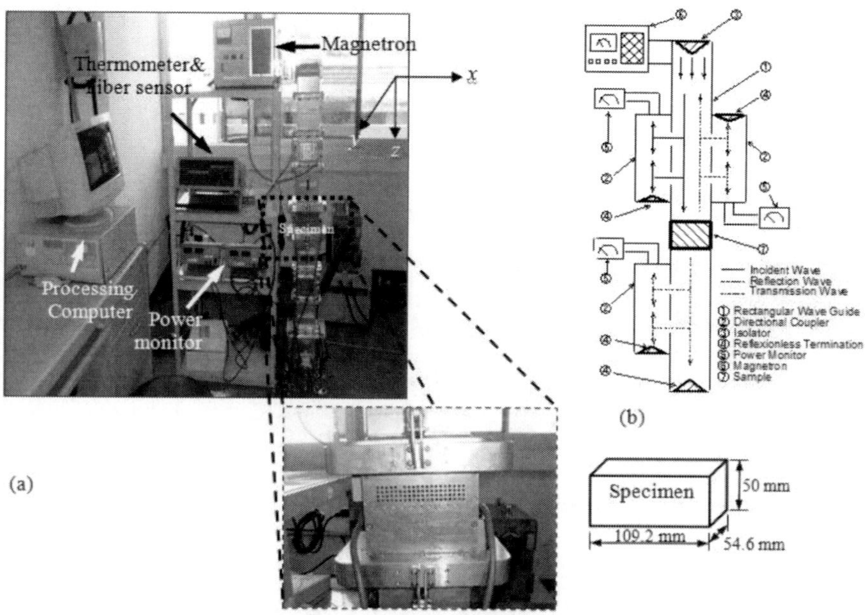

Figure 4.1. (Continued).

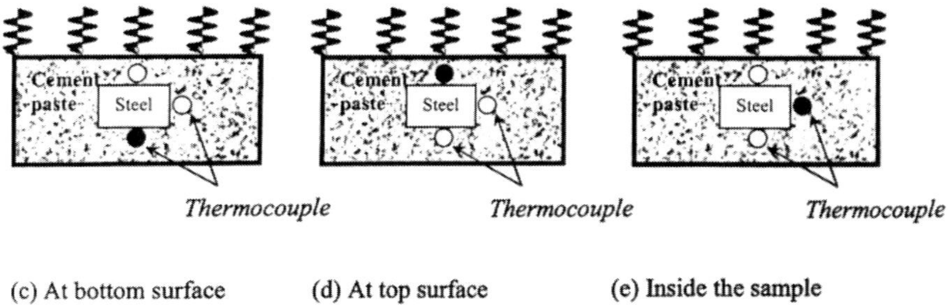

Figure 4.1. (a) Experimental set up and (b) schematic showing direction of microwave components (Incident wave, reflected wave, and transmission wave), and (c), (d), (e) are the positions of thermocouples.

For the specimens subjected to microwave curing, the delay times at 30 min after mixing, at the initial setting time, and at the final setting time are determined by the present study for microwave application at the following power levels and durations: power levels of 50, 100, and 150 watt and times of 0, 10, 20, and 30 min. Furthermore, the decelerated period used normal cooling and vacuum conditions.

4.3.5. Equipment and Procedures

4.3.5.1. Scanning Electron Microscope (SEM)

A Scanning Electron Microscope (SEM), specifically an International Scientific Instruments ISI-130 electron microscope, was used to determine the microstructure and morphology of the samples. A dual-stage, dual-screen microscope, it had five lenses and used an energy x-ray (EDX) analysis system. The maximum practical resolution was approximately 100,000 times. The specimens were glued onto a sample stub using carbon tape and then placed in a vacuum chamber and sputtered with gold for approximately 40 min at a 75 voltage. The gold film created a route by which the electrons could be conducted off the surface of the sample; otherwise, the accumulation of electrons on the sample surface would have led to charging and a fuzzy picture. The image was displayed on two 30 cm cathode ray tubes.

4.3.5.2. Microscopes: Powder X-ray Diffraction (XRD)

The crystalline phase identification of the various samples was performed on a Scintag X-ray Diffractometer. This diffractometer is equipped with a copper target X-ray source, monochromator, and Tl-drifted NaI scintillation detector. Dried-powder samples were packed into a cavity of a zero-background quartz slide and placed on a goniometry. Most of the subsequent scans were taken from 0 to 70^0 2Θ at a rate of 2^0 2Θ per min. The diffractometer is controlled by a VAX 3100 mini-computer and contains a matched software that can be used to display the data on the computer screen and through which data manipulation can be performed. The raw data can be downloaded into a personal computer and the results printed on either a Hewlett Packard plotter or a laser printer. This software package can distinguish between amorphous and crystalline peaks and between $K_{\alpha 1}$ and $K_{\alpha 12}$ peaks. In all cases, the raw data used contained some information pertaining to amorphous phases, in the form of humps in the spectra, which would otherwise have been stripped away by the software.

4.3.5.3. Thermal Analysis

Differential thermal analysis of a few selected samples was conducted using a NETZSCH, STA449C Jupiter© analyzers (Sample temperature: 0.1°C resolution, and better than 0.5°C accuracy, Temperature range: room temperature to 1,550°C and Heating rate: 0.1-99.9°C/min. In all cases 40 – 50 mg of the sample were used, and alumina was thc reference material. The heating rate was 10°C/min, and the atmosphere was air. The instrument was suitably adjusted in some cases to obtain the best thermograms in the temperature range of 30 to 600°C.

4.3.5.4. Axial and Unconfined Load Compressive Strength

The respective compressive strengths of the cement pastes, mortars, and concrete samples were tested using a compressive strength apparatus in accordance with the ASTM C 39 at 8 h, 24 h, 3 days, and 28 days (Figure 4.2).

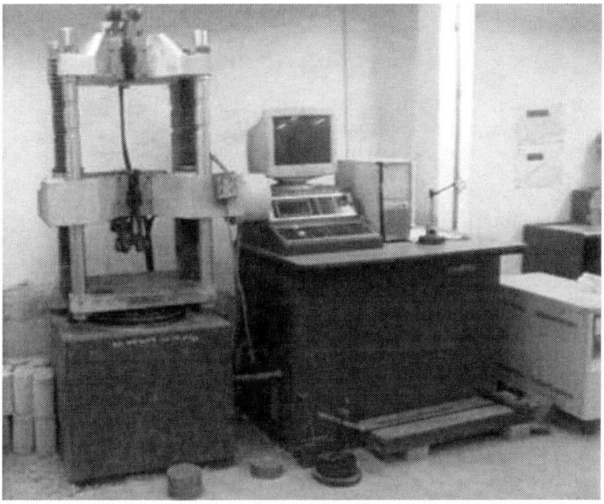

Figure 4.2. Universal testing machine.

4.4. Results and Discussion

4.4.1. Influence of the Mix Proportions of Reinforced Cementitious Materials

4.4.1.1. Water-Cement Ratios (c/s)

The effect of water-to-solid ratios on microstructures characteristics of the reinforced pastes subjected microwave energy is presented. These characteristic is investigated by scanning electron microscope (SEM) with dispersive X-ray (EDX), X-ray diffraction (XRD) and thermal analysis. The pastes used were proportioned at w/c ratios of 0.25, 0.38 and 0.45. After mixing and molding, they were cured at room temperature by wrapping with polyethylene plastic until the delay time (time after mixing until introducing microwave energy with a single-mode cavity) for 30 min The temperature profile and microwave power of 100 watt with a specific application time 30 min of the pastes is shown in Figure 4.3. With limitation of the microwave equipment used, it can only adjust manually the steps of microwave power level such as 50, 100, 150, 200 ,…, 3000 watt, therefore the optimal microwave levels with the aforementioned research should be set at 100 ± 5 watt.

Temperature profiles obtained from averaging the five monitored data at bottom surface (Figure 4.3(a)), top surface (Figure 4.3(b)) and at the middle (Figure 4.3(c)). The temperatures increase monotonically among the positions of measurement during the microwave curing process and reach a maximum of 105°C at the bottom surface of the cured cement paste. Significantly, the paste at lower water-to-cement ratio experiences high temperature rise, or 1CW/C_P0.25 > 1CW/C_P0.38 > 1CW/C_P0.45. This is because of two inclusive effects; (i) heat liberation from hydration reaction is increased as low content of water in the system, and (ii) heat from interaction between the microwave energy and internal water leading to superposition of them. In other words, the addition heat from microwave energy can change the kinetics of hydration in accordance with Arrhenius's law.

When comparison the obtained temperatures at three positions, the bottom of the cured specimen has highest temperature levels than those of other sides. It may be due to that fact that during temperature rise, the water at the top side of specimen may evaporate, so the temperature was dropping gradually, while at the bottom the evaporation of water is difficult. As a result the heat accumulates on the specific side providing temperature increase with high rate than the other sides.

The typical micrographs of the 1CW/C_P0.38 paste at the age of 4 h after mixing, 28 days after curing in lime-saturated deionized water, and subjected to microwave energy are shown in Figure 4.4. It is clearly seen from Figure 4.4(a) and 4.4(b) that the samples consist of hydrated phases and pores, as well as cores of $Ca(OH)_2$ dendrite crystals or other crystals (marked CH), calcium silicate hydrate (C-S-H), and granular structure. Furthermore, some ettringite (Aft) is found in case of specimens were cured by microwave energy. It can be described that in the early stages of reaction of the 27°C sample, very small (about 1 μm) irregularly-shaped ettringite was formed; but at the same curing time, needle-like ettringite had already formed in 60 °C samples.

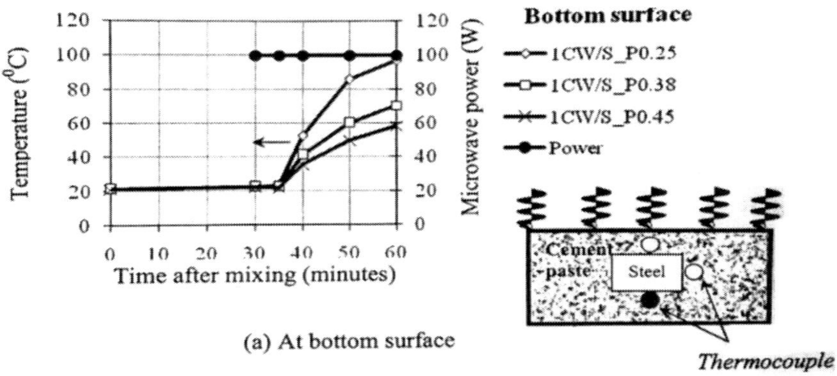

(a) At bottom surface

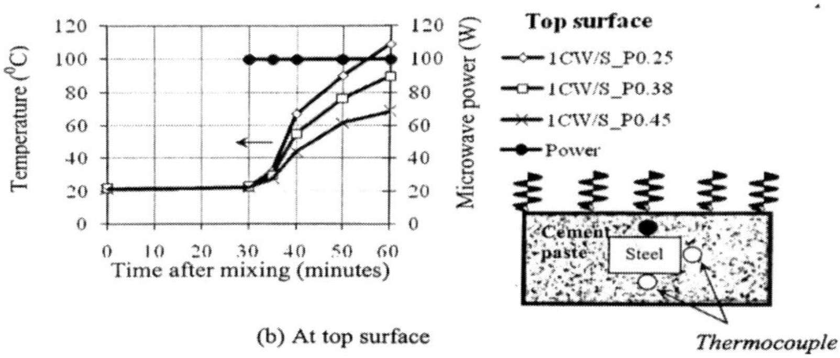

(b) At top surface

Figure 4.3. (Continued).

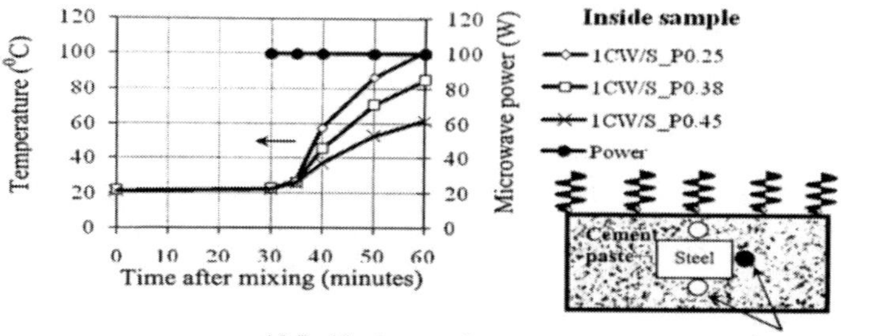

(c) Inside the sample

Figure 4.3. Temperature and power history during applying microwave energy of various cement pastes with different water-to-solid mass ratios.

(a) 1CW/C_P0.25 at the age of 4 h.

(b) 1CW/C_P0.38at the age of 4 h.

(c) 1CW/C_P0.25at the age of 28 days.

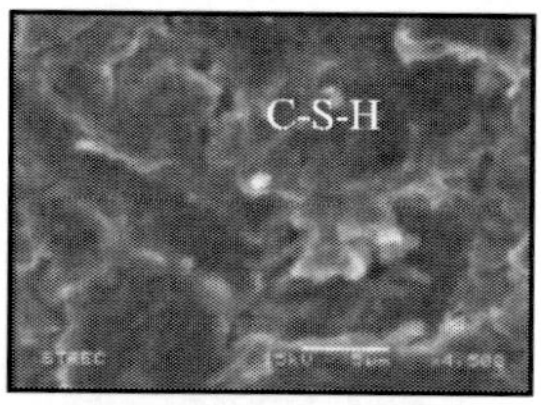

(d) 1CW/C_P0.38 at the age of 28 days.

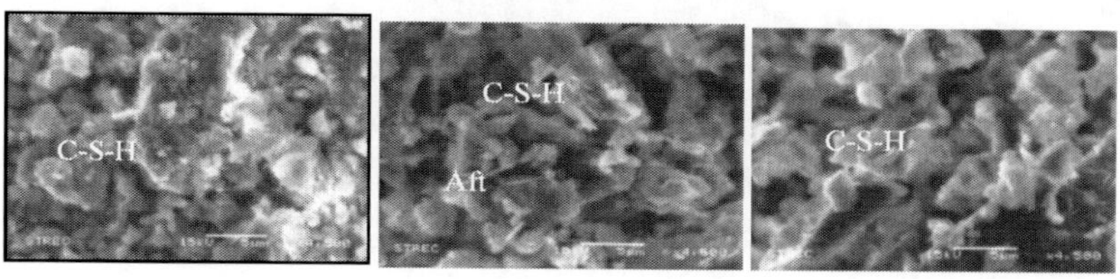

(e) 1CW/C_P0.25 after applying MW energy power 100 W for 30 min.

(f) 1CW/C_P0.38 applying MW energy power 100 W for 30 min.

(g) 1CW/C_P0.45 applying MW energy power 100 W for 30 min.

Figure 4.4. Micrographs of various cement pastes with different water-to-solid ratios at 4 h subjected to lime-saturated deionized water curing and microwave energy.

The composition of the calcium silicate hydrate (C-S-H) gel and other products formed in hydrated Portland cement systems has been studied in terms of its Ca/Si and Al/Ca ratios by several researchers. The results reported vary due to the individual characteristics of the cements hydrated and the conditions of hydration, as well as the experimental technique used for the characterization. Calculated Ca/Si ratios also can be obtained from the measured CH concentration and the degree of hydration of the calcium silicates. Cement hydration products are commonly classified as two products. The former are those hydration products formed within the original cement grain boundaries, and the latter are those products formed in the space initially occupied by water. Such definitions are used in this work. In the products, the main cement hydration product is C-S-H gel, which is principally responsible for the properties of the hydrated cement.

Spectrums of various cement pastes with different water-to-cement ratios at 4 h subjected to normal curing (lime-saturated deionized water) and microwave energy are shown in Figure 4.5. However, the results in the form of atomic (molar) ratios of several elements with respect to calcium, from the EDS analysis composition are also represented Figure 4.6. It shows plots of the Al/Ca ratio versus the Si/Ca ratio for the pastes hydrated at microwave power 390 watt for 45 min. The (X) corresponds to the 0.45-w/c paste analyses, as the hydration temperature decreases from 100° to 62°C, the Al/Ca ratio changes from 0.022 to 0.091.

From the SEM–EDS results it was observed, although the measured Ca/Si ratios of the pastes were similar in magnitude that they consistently decreased when the temperature was decreased from 100° to 62°C. In this study, the same SEM equipment was always used and an effort was made to perform all the analyses under essentially the same experimental conditions. Hence, the trends are believed to be reliable, even if the absolute values are not. Some results in the literature agree with this observed trend, but some others differ. Sorption of sulfate ions by C-S-H gel, as discussed below, may require extra calcium ions for charge balance, hence providing a mechanism for decreasing in this ratio. The observed decrease in the Ca/Si ratio was, however, small and similar in magnitude to the standard deviation of the results. In an independent investigation, the difference between the Ca/Si ratios, this ratio and its variation with curing temperature may vary with the extent of solid solution of foreign ions incorporated in the C-S-H gel and the amount of other phases, notably CH, intermixed in the gel phase. It seems prudent to conclude overall that, until analyses can be obtained with lower errors, the Ca/Si ratio does not vary in a systematic way with an increase in the curing temperature. It also has been observed that, for the cement pastes cured at the highest temperature, more sulfur and less aluminum are retained in the pastes.

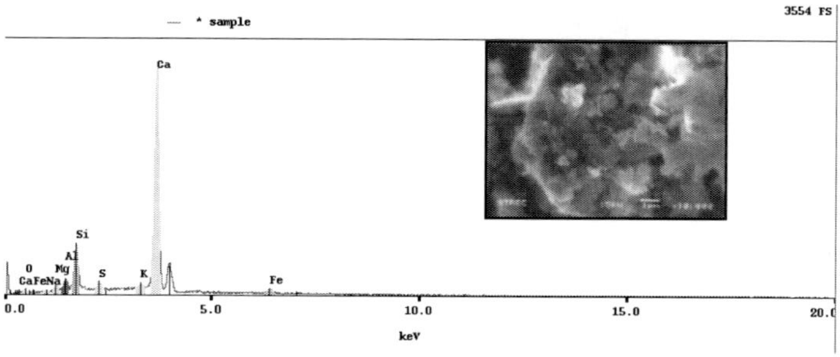

(a) 1CW/C_P0.25 after applying a microwave power 100 watt for 30 min.

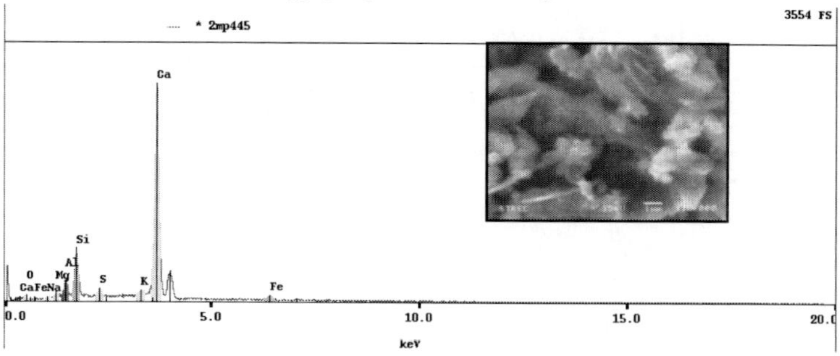

(b) 1CW/C_P0.38 after applying microwave power 100 watt for 30 min.

Figure 4.5. (Continued).

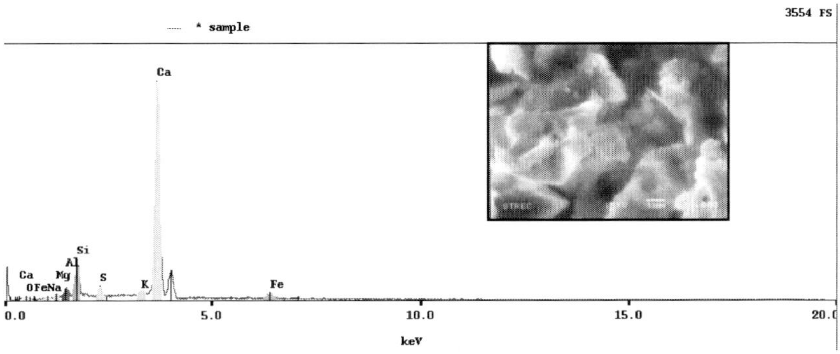

(c) 1CW/C_P0.45 after applying microwave power 100 watt for 30 min.

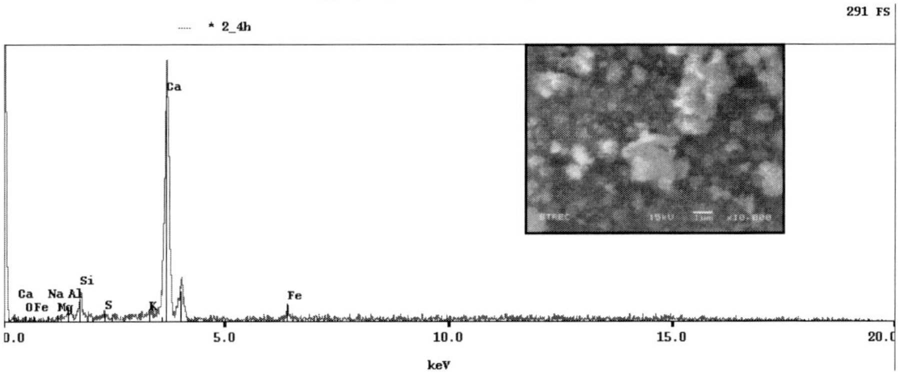

(d) 1CW/C_P0.38 at the age of 4 h.

Figure 4.5. Spectrums of various cement pastes with different water-to-solid ratios at 4 h subjected to normal curing (lime-saturated deionized water) and microwave energy.

X-ray diffractometry was used to determine the degree of crystallinity of the hydrated cement products and the existence of crystalline coexisting phases. Figure 4.7 shows x-ray patterns of the hydrated products in the pastes of 1CW/C_P0.38 after applying microwave power 100 watt for 30 min. The phases identified include calcium silicate hydrate (Ca_3SiO_5), calcium hydroxide ($Ca(OH)_2$), residual lime (CaO) and Xenotile ($Ca_6(SiO_3)_6(H_2O)$).

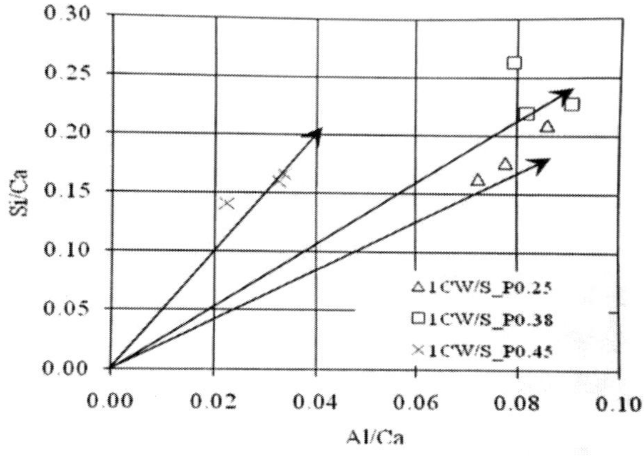

Figure 4.6. Atom ratio of Si/Ca versus Al/Ca of the pastes of 1CW/C_P0.25, 1CW/C_P0.38 and 1CW/C_P0.45 after applying microwave energy power 100 watt for 30 min.

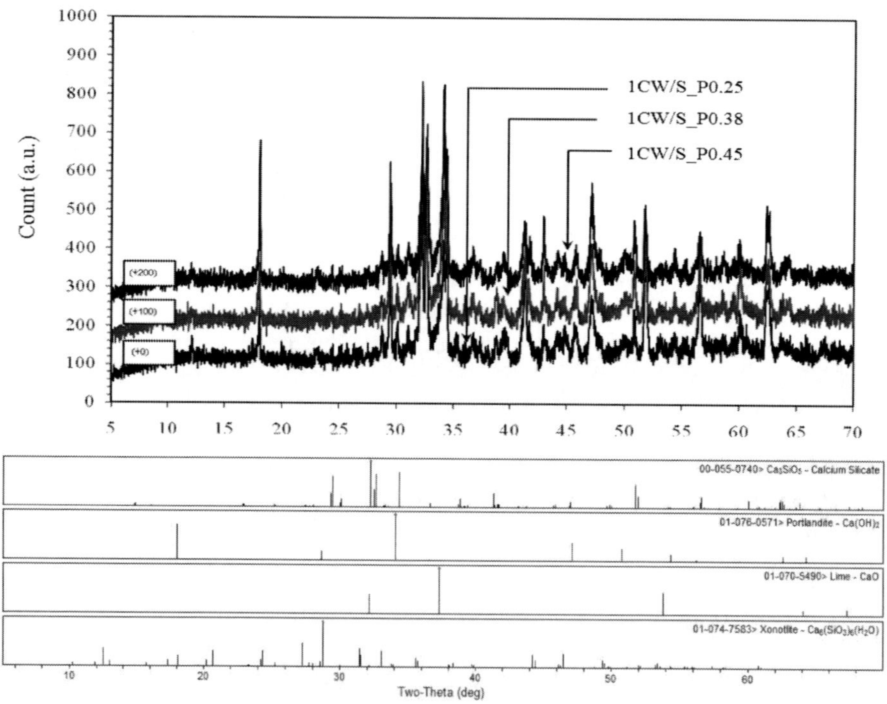

Figure 4.7. X-ray diffraction of the pastes of 1CW/C_P0.25, 1CW/C_P0.38 and 1CW/C_P0.45 after applying MW power 100 watt for 30 min.

Regarding the effect of water-cement ratios on the phase characteristics of 1CW/C_P0.25, 1CW/C_P0.38 and 1CW/C_P0.45 after applying microwave power of 100 watt for 30 min as shown in Figure 4.7, it is found that the calcium silicate hydrate (Ca_3SiO_5), calcium hydroxide ($Ca(OH)_2$) phases are similar.

Compressive strengths of the pastes of 1CW/C_P0.25, 1CW/C_P0.38 and 1CW/C_P0.45 after applying microwave power of 390 watt for 45 min compared to normal curing are shown in Figure 4.8. It was found that when cured at elevated temperatures, pastes develop strength quite rapidly. At the age of 8 h after microwave curing at 100°C (at middle) 1CW/C_P0.25 paste attained a strength of 25.1 MPa (239% higher than the normal-cured paste); at 24 h curing the strength was 68.3 MPa, and 7 and 28 days the strength were 74.9 and 75.2 MPa, respectively. For the elevated temperature curing with 1CW/C_P0.38 and 1CW/C_P0.45 pastes, the strength obtained at 8 h was 10.9 MPa, exceeding the strength of the paste cured without microwave energy. However, at 24 h and later ages the strength of the pastes cured at elevated temperature was 23 to 46% lower than that of the paste cured at room temperature. This is due to decrease of the water-to-cement ratio less than about 0.40. According to Taylor, when water-to-cement ratio should equals to 0.40, all the water molecules exist as gel and combined water in C-S-H structure; that is, there is no capillary water in the case of full hydration taking place. However, some capillary pores should be preserved as a path, through which water molecules can come in and react with cement grains near the pores. Meanwhile, these pores could serve as a space for gel expansion. The existence of a few capillary pores might be favorable for the structure and strength of the hardened cement paste. The above statement is consistent with our experimental results; that is if w/c or w/s ratio is less than 0.40, it is difficult for the full hydration of cement and unfavorable for the strength of concrete.

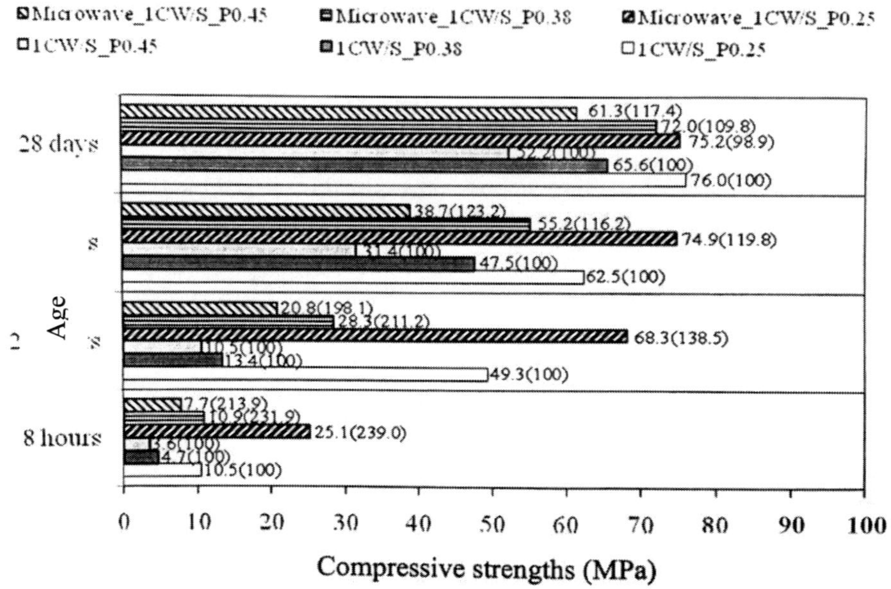

Figure 4.8. Compressive strengths of the pastes of 1CW/C_P0.25, 1CW/C_P0.38 and 1CW/C_P0.45 after applying microwave power 390 watt for 45 min compared to normal curing.

The evaluation of the hydration rate in the microwave -cured and normal-cured cement pastes was based on the Thermogravimetry (TG) curves and consisted of the following steps: (a) determining the total combined water: this is the total water incorporated in the cement paste, which corresponds directly to weight loss of up to 550°C; (b) determining the amount of $Ca(OH)_2$, which is found by weight loss in the range of 400–550°C, converted to $Ca(OH)_2$, and then added to the carbonated $Ca(OH)_2$; and (c) determining the amount of water combined in the hydration products (other than calcium hydroxide), which corresponds directly to weight loss of up to 300°C. Any changes in the water content of the hydration products indicate that these compounds have changed in terms of how they are constituted and/or their stoichiometry.

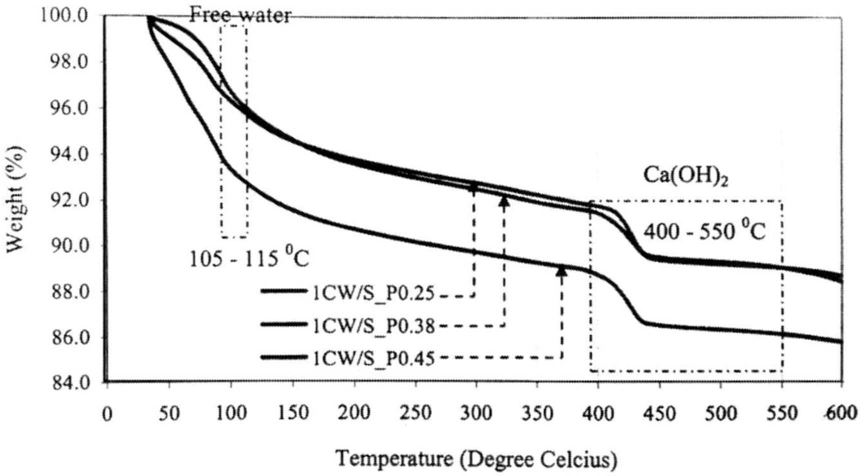

Figure 4.9. Thermogravimetric analysis (TG) by comparison among different temperatures and weights.

Figure 4.9 presents the weight loss that accrues from the water lost in the hydration reactions, the loss of $Ca(OH)_2$ content, and combined loss of water and $Ca(OH)_2$ content for the cement pastes when subjected to microwave energy of 390 watt for 45 min. It can be seen that the cement paste at w/c = 0.25, and that the total water, total $Ca(OH)_2$ content, and water and $Ca(OH)_2$ combined are 10.990%, 2.426%, and 4.514%, respectively. For the 0.38 paste, the corresponding figures are 10.998%, 2.723%, and 7.243%, respectively; for the 0.45 paste the figures are 13.889%, 2.587%, and 10.307% respectively. For $Ca(OH)_2$ content, the 0.38 w/c MW-cured paste is quite different from the others because this w/c is optimal for reacting to fully hydrated products, while the higher w/c pastes produce a low concentration of Ca^{2+} and OH^- and consequent to crystallize a low $Ca(OH)_2$ content (Taylor, 1997). Apart from $Ca(OH)_2$ content, for the case in which water was combined with $Ca(OH)_,$ it is also similar to total water content in the hydration products.

In summary, when microwave power and time of application were kept, the relative water content of the pastes' hydration products were affected by different w/c, while their $Ca(OH)_2$ contents were affected due to the completion of hydration reaction which is occurred at the water-cement mass ratio (w/c) of 0.38 (Taylor, 1986), while a lower w/c is not sufficient ions (Ca^{2+} and OH^-) quantities, and a higher w/c is not concentration to crystalline.

4.4.1.2. Pozzolan Materials in Reinforced Cement Paste

The effects of microwave energy used with a multi-mode system on pozzolan materials are discussed in this section. Fly ash (FA), matakaolin (MK), and silica fume (SF) were taken into account. The pastes containing pozzolan materials were proportioned at w/c ratios of 0.38 and were also replaced at 20% by weight. After mixing and molding, they were wrapped in polyethylene plastic in order to be cured at room temperature with a delay time after mixing for 30 min. The temperature profile of the pastes at microwave power of 100 watt with a specific application time of 30 min is shown in Figure 4.10. The temperature profiles were obtained by averaging the five monitored data at the top surface (Figure 4.10(a)), at the bottom surface (Figure 4.10(b)) and in the inside (Figure 4.10(c)). The temperature increases monotonically among the points of measurement during the microwave-curing process and reached a maximum temperature of 89°C at the bottom surface of the microwave-cured FA–cement paste, 72.9°C at the bottom surface of the microwave-cured MK–cement paste, and 62.4°C at the bottom surface of the microwave-cured SF–cement paste. Significantly, for pastes containing pozzolan temperature rise relates to the free water that remains the pozzolan has been absorbed, and the amount of heat that microwave energy generates is also decreased as well. It is well-known that the FA particles are more rounded in shape than are the MK and SF particles, and as a result the FA–cement paste shows the highest temperature increase of the three.

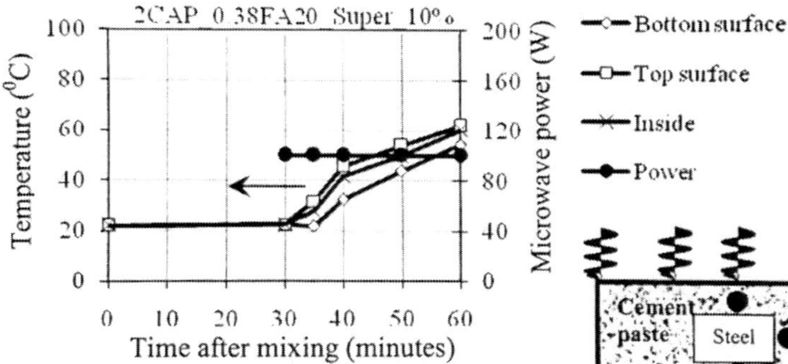

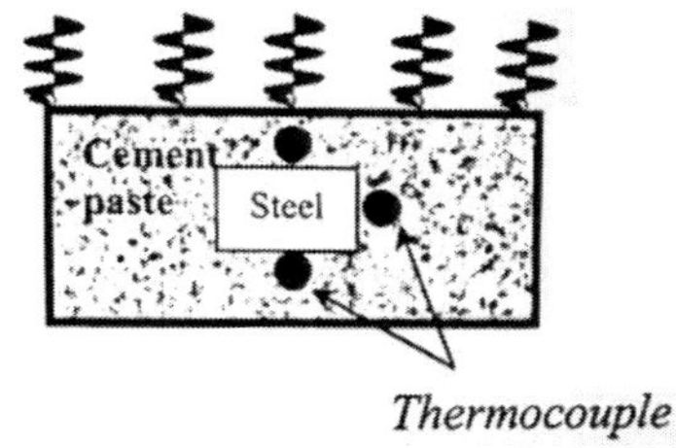

(a) Pulverized fuel ash (FA)

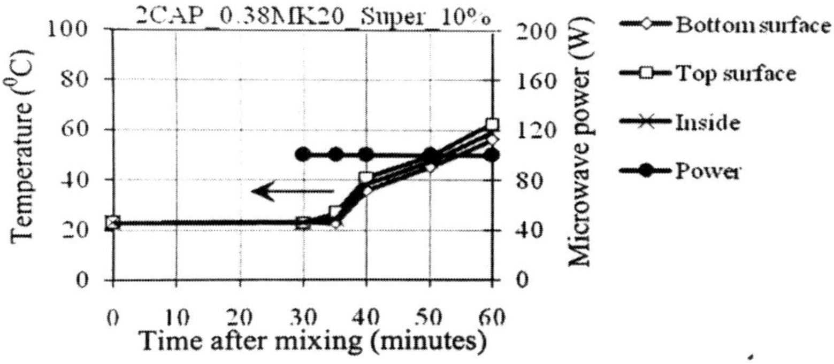

(b) Metakaolin (MK)

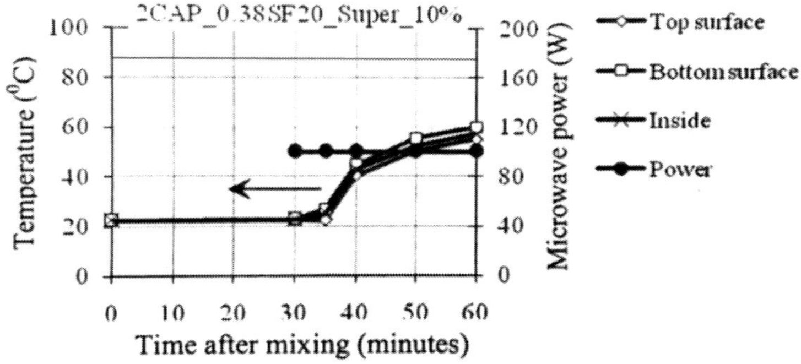

(c) Silica fume (SF)

Figure 4.10. Temperature and power history during applying microwave energy of various cement pastes mixed with pozzolan materials.

Furthermore, at early-age microwave curing, the rate at which the temperature increases is higher than the sequential temperature profile of the FA–cement paste increases. This is because the application of microwave energy accelerates rapidly the hydration and pozzolan reactions and so affects the production of Portlandite ($Ca(OH)_2$). This is subsequent to an increase in the concentration of Portlandite, which, in turn, also speeds up the pozzolanic reaction. And, it should be noted that an increase in this reaction is accompanied by an increase in heat liberation or temperature rise.

Similar to the effects noted in the previous section, the bottom of the microwave-cured paste has higher temperature levels than do the top side or the inside. At the top side, the temperature level was affected by the rate at which the water evaporated of water; that is, the temperature dropped gradually; at the bottom water evaporation took place even more slowly. While the heat accumulated on the bottom side, and thus caused the temperature to increase more quickly at the bottom than at the other points of measurement.

The typical micrographs of FA–cement, MK–cement, and SF–cement pastes at the age of 4 h after mixing and curing by wrapping with a plastic sheet; at 28 days after curing in lime-saturated deionized water; and after being subjected to microwave energy at a microwave power of 100 watt for 30 min are shown in Figure 4.11. The FA–cement paste shows little difference among the microstructures at the three detecting times. Some small fibers appear at the age of 4 h, but none appear in the FA–cement paste that was heated by microwave energy. The MK–cement paste, shows little difference between the normal-cured and microwave-cured pastes. On the other hand, for the SF–cement paste the SF particles under microwave curing have dispersed more than with the 4-h normal curing, while the paste at 28 days looks like a plate of C–S–H in which SF particles cannot be distinguished.

(a) 2CAP_0.38FA20_ Super_10% at the age of 4 h

(b) 2CAP_0.38MK20_ Super_10% at the age of 4 h

(c) 2CAP_0.38SF20_ Super_10% at the age of 4 h

(d) 2CAP_0.38FA20_ Super_10% at the age of 28 days.

(e) 2CAP_0.38MK20_ Super_10% at the age of 28 days.

(f) 2CAP_0.38SF20_ Super_10% at the age of 28 days.

(g) 2CAP_0.38FA20_ Super_1% after applying microwave energy power 100 watt for 30 min.

(h) 2CAP_0.38MK20_ Super_1% after applying microwave energy power 100 watt for 30 min.

(i) 2CAP_0.38SF20_ Super_ 1% after applying microwave energy power 100 watt for 30 min.

Figure 4.11. Micrographs of various cement pastes mixed with pozzolan materials with different water-to-cementitious ratios at 4 h subjected to normal curing (lime-saturated deionized water) and microwave energy.

Figure 4.12 shows the atom ratio of Si/Ca versus Al/Ca for the pastes after microwave power of 390 watt had been applied for 45 min. It can be seen that in the case of the normal paste (×), the cluster of Si/Ca versus Al/Ca takes place in the narrow range of the Si/Ca ratios equal to 0.220–0.263 while the Al/Ca ratios equal to 0.079–0.091. The paste mixed with FA (CAP_0.38FA20_Super_10%,) has a wider dispersion of Si/Ca versus Al/Ca than does the normal paste. In the paste mixed with MK (2CAP_0.38MK20_Super_10%), the Si/Ca versus Al/Ca has shifted to achieve a higher Al/Ca ratio in comparison with the normal paste. In the SF paste (2CAP_0.38SF20_Super_10%), the Si/Ca versus Al/Ca shifted to achieve a higher increase in the Si/Ca ratio in comparison with the normal paste. These phenomena can be explained by noting that the chemical composition of FA shows a higher aluminum oxide content than does Portland cement and the effect of this composition is higher than that of the MK–paste. On the other hand, SF is a silica-rich material; therefore, the Si/Ca ratios in the SF–paste are shifted from 0.582 to 1.184. It appears, therefore, that microwave energy can expand the dispersion of silica in an SF paste. Nevertheless, this dispersion has a wide range, that of 0.563 to 1.163.

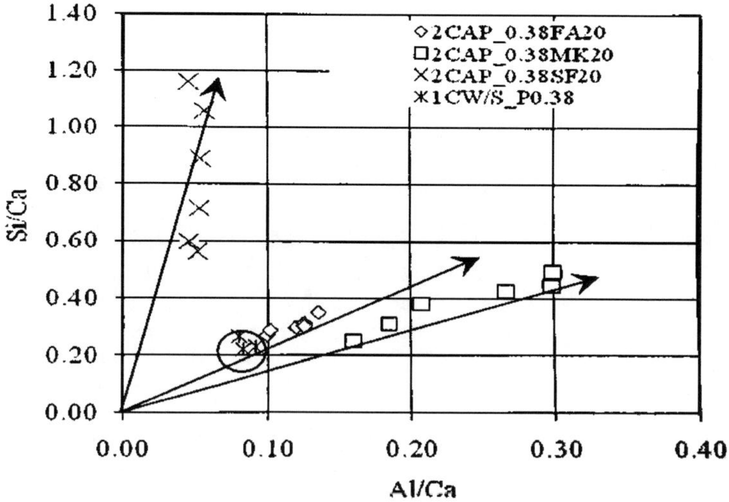

Figure 4.12. Atom ratio of Si/Ca versus Al/Ca for the pastes of 2AP_0.38FA20, 2CAP_0.38MK20 and 2CAP_0.38SF20 after applying microwave power 100 watt for 30 min.

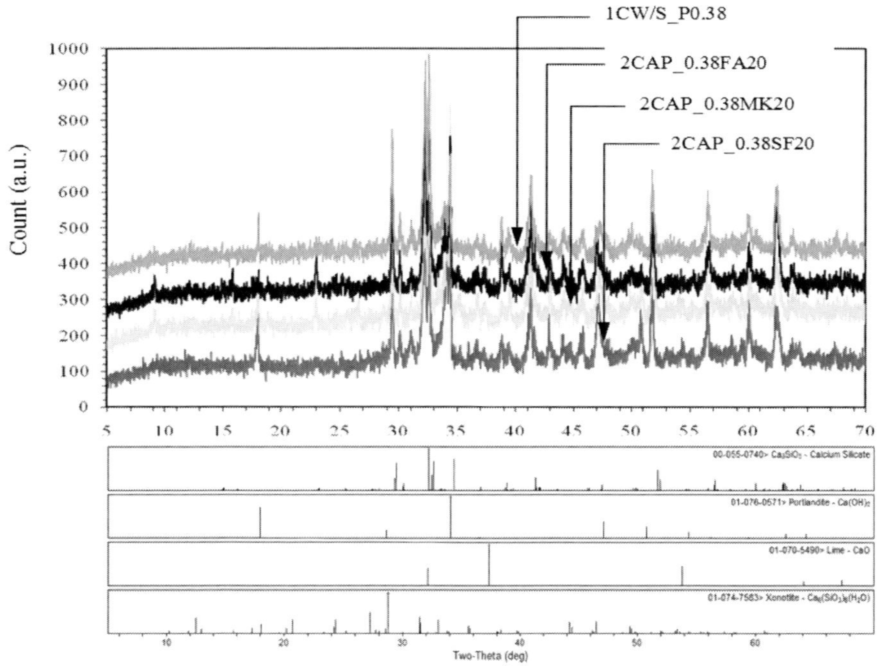

Figure 4.13. X-ray diffraction of 2CAP_0.38FA20, 2CAP_0.38MK20 and 2CAP_0.38SF20 after applying microwave power 100 watt for 30 min.

X-ray diffractometry was used to determine the degree of crystallinity of the hydrated cement products and the existence of crystalline coexisting phases. Figure 4.13 shows X-ray diffraction of CAP_0.38FA20_Super_1%, 2CAP_0.38MK20_Super_1% and 2CAP_0.38SF20_Super_1% after MW power of 390 watt had been applied for 45 min. The phases identified include calcium silicate hydrate (Ca_3SiO_5), calcium hydroxide ($Ca(OH)_2$), residual lime (CaO), and Xenotile ($Ca_6(SiO_3)_6(H_2O)$).

As shown in Figure 4.13, in regard to the effect of pozzolan material on the phase characteristics of CAP_0.38FA20_Super_1%, 2CAP_0.38MK20_Super_1%, and 2CAP_0.38SF20_Super_1% after microwave power of 390 watt had been applied for 45 min, the calcium silicate hydrate (Ca_3SiO_5) and calcium hydroxide ($Ca(OH)_2$) phases are similar.

The compressive strength of the CAP_0.38FA20_Super_1%, 2CAP_0.38MK20_Super_1%, and 2CAP_0.38SF20_Super_1% pastes after the application of microwave power of 376 watt for 45 min compared to compressive strength under normal curing conditions are shown in Figure 4.14. It was found that when cured at elevated temperatures, the pastes developed strength quite rapidly, especially the SF–cement paste. At the age of 8 h after microwave curing the FA–cement paste gained a compressive strength of 4.1 MPa (41000% higher than that of the normal-cured paste); at 24 h curing, FA–cement pastes showed a compressive strength of 9.2 MPa, and at 7 and 28 days the strengths were 25.1 and 40.6 MPa, respectively. For the elevated temperature curing with the MK– and SF–cement pastes, the strength obtained at 8 h was 8.6 and 12.1 MPa, which exceeded the strength of the non-microwave-cured pastes. The pastes containing pozzolan material showed strength development behaviors that differed from those of the normal paste; that is it can develop continuously the strength after 24 h and later ages. This is due to the pozzolan reaction, which produces the secondary C–S–H content and the hydration reaction (Massazza, 2003); these effects contribute to the heat generated by the microwave and the heat liberated from hydration which, in turn, contribute to the later-age strength development of the pozzolan pastes.

The compressive strength development of the pozzolan pastes relate to their $Ca(OH)_2$ content: that is, strength development is related to the weight loss of the sample in the range of 400–550°C, the weight loss is converted to $Ca(OH)_2$ and then added to the $Ca(OH)_2$ that has been carbonated. Figure 4.15 shows that the $Ca(OH)_2$ content of the FA–, MK–, and SF–cement pastes were 3.08, 2.36, and 1.59 (%wt.), respectively. This illustrates that the that the SF–cement paste consumed more $Ca(OH)_2$ in the pozzolan reaction and produces more C–S–H, the combination of which means that of the three pastes, this one achieves the highest compressive strength.

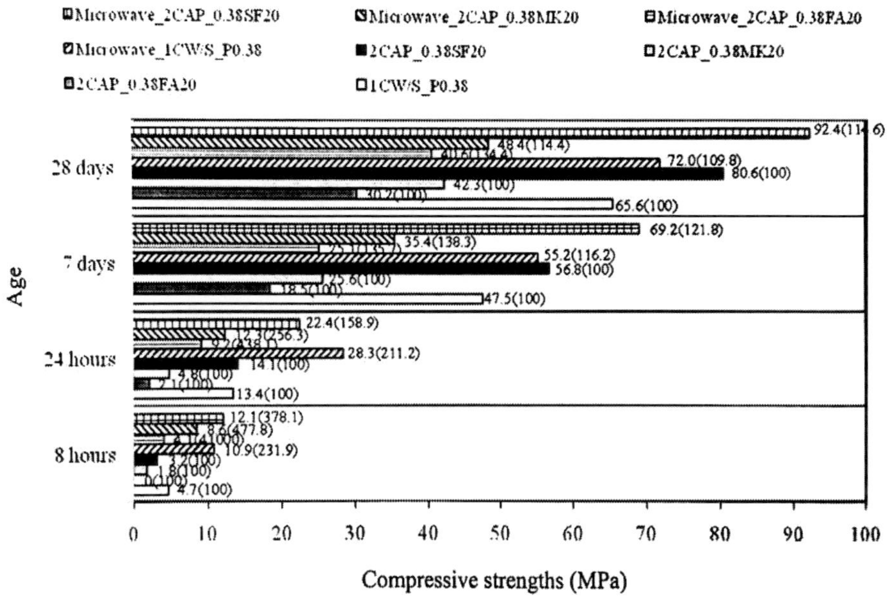

Figure 4.14. Compressive strengths of the pastes of CAP_0.38FA20, 2CAP_0.38MK20 and 2CAP_0.38SF20 after applying microwave power 100 watt for 30 min compared to normal curing.

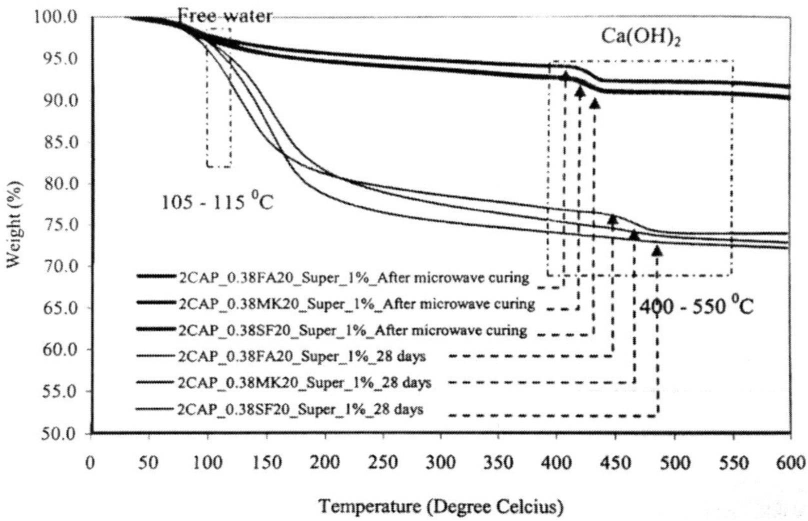

Figure 4.15. Thermal analysis results by comparison among different temperatures and weights of the pastes with/without pozzolan materials.

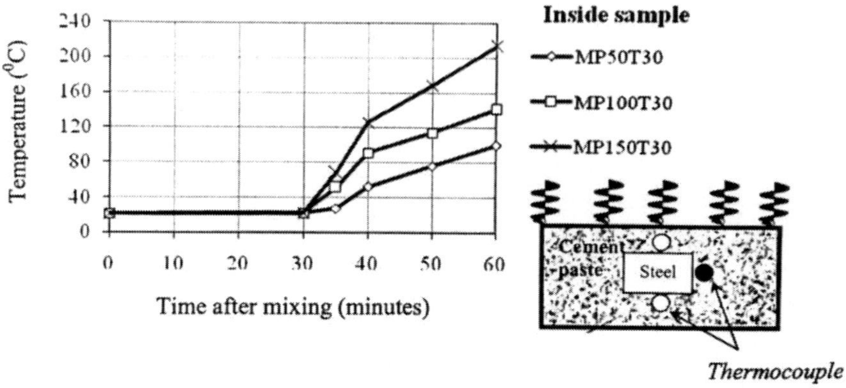

Figure 4.16. Temperature and power history during the application of microwave energy to an inside of the mortar sample (3CAM0.38_1:2.75) with different microwave power levels.

4.4.1.3. Reinforced Mortar and Concrete

Figure 4.16 shows the temperature increases and microwave power used throughout the curing time of the mortar (3CAM0.38_1:2.75) proportioned with Portland cement Type I: Ottawa sand at 1: 2.75 by weight and water-cement ratio of 0.38. The temperatures increases are related to increases in of microwave power. This is because the heat generated by microwave energy varies in correspondence with the electric field intensity ($\vec{E}$), as shown in Eq. (2.19). In other words, when the microwave power increases, so too does the intensity of the electric field. Moreover, the temperature behaves similarly for all the monitored points. In terms of dielectric behavior, the SiO_2 in sand does not absorb microwave energy; therefore, the heat that accrues from microwave energy proceeds from hydration of the cement particles and water molecules.

Regarding the rate at which the temperature increases, in the first period during which microwave is introduced inwards to the microwave-cured mortar, the rate of increase is higher than that of the later period. This is because during the first period, the abundance of water inside the mortar consistently affects increases in the amount of heat generated. On the other hand, during the later period the water's composition changes to C–S–H and is evaporated outward, which causes a gradual reduction in the amount of heat generated as the amount of available water, especially free water, decreases.

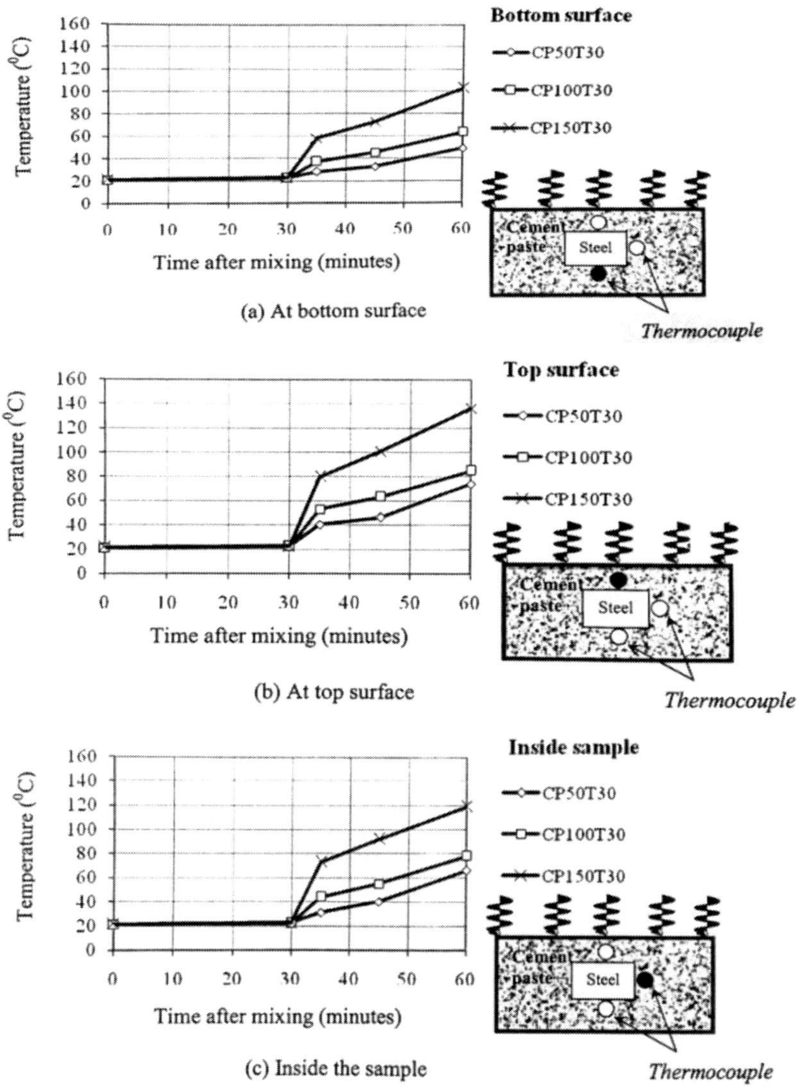

Figure 4.17. Temperature and power history during the application of microwave energy to concrete (3CAC0.38_1:1:1) with different microwave power levels.

It was found that the rate at which temperatures increases at the end of curing, corresponds to the power level of the microwave energy applied. For example, at the top surface where water evaporation take place, the final temperature at microwave power levels 50, 100, and 150 watt run at 69°C, 123°C, and 190°C, respectively. This shows that though the microwave power level was increased up to two times, the temperature does not raise a half of the microwave power level. This high rate of heat generation at high power levels directly increases more rapidly the rate at which the temperature of the heated mortar increases. In addition, the temperature as immediately generated can accelerate hydration; that is, this reaction also generates heat.

A comparison of the differences in the temperature increases of the microwave-cured paste (Figure 4.16) and the microwave-cured mortar at the same water-cement mass ratio shows that the mortar stores less energy than does the paste. This is due to high specific heat of sand (664 J/kg-°C), which causes the temperature of mortar rises up at a lower rate. Furthermore, as water evaporates from the cement paste, its temperature decreases more rapidly than does the of the mortar sample.

Figure 4.17 shows the temperature and power history during the application of microwave energy to a concrete (3CAC0.38_1:1:1) with different microwave power levels. The concrete is proportioned as follows: Portland cement: Ottawa sand: crushed limestone rock equal to 1:1:1. This temperature behavior differs in clear ways from the temperature behavior observed in the mortar. In detail, during the early period of microwave curing the temperature increases rapidly; that is, it rises by an average of every 5 min. Subsequently, the temperature increases at a constant rate. All the monitored points—the top surface, bottom surface, and inside—behave similarly in this regard, which indicates that the presence of crushed lime stone rock affects the rate at which the temperature increases as well as the final temperature achieved. This may be because crushed limestone rock has a rough surface to which it is easier for water molecules to attach than the smoother surface of Ottawa sand. This result shows that the heat generated within this sample is lower than that of is generated in the mortar sample. This means that when microwave energy is introduced to the concrete sample, the sample maintains temperature or absorbs heat better than does the mortar under similar conditions.

Typical micrographs of the mortar (3CAM0.38_1:2.75) subjected to microwave energy, and concrete (3CAC0.38_1:1:1) subjected to microwave energy at 50 watt for 30 min are shown in Figure 4.18. It is not clearly seen from Figure 4.18(a) and 4.18 (f) that the samples consist of hydrated phases and pores. On its surface, the large particles appear to be agglomerated with a small particle, especially in the microwave-cured concrete.

a) 3CAM0.38_1:2.75 at a microwave power 50 watt.

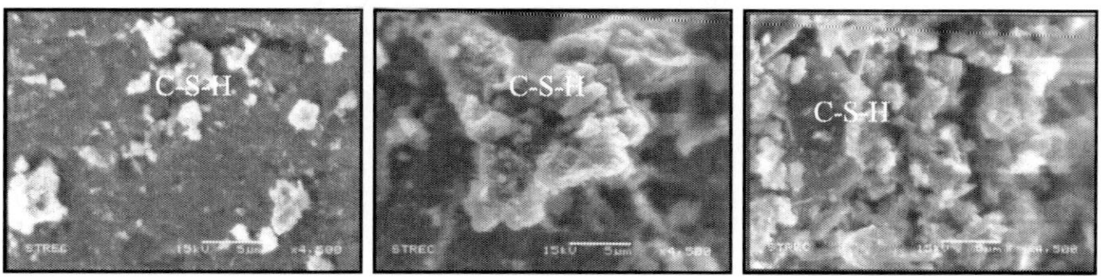

(b) 3CAC0.38_1:1:1 at microwave power 50 watt.

(c) 3CAC0.38_1:1:1 at microwave power 100 watt.

(d) 3CAC0.38_1:1:1 at microwave power 150 watt.

Figure 4.18. Micrographs of mortar and concrete with different microwave power levels.

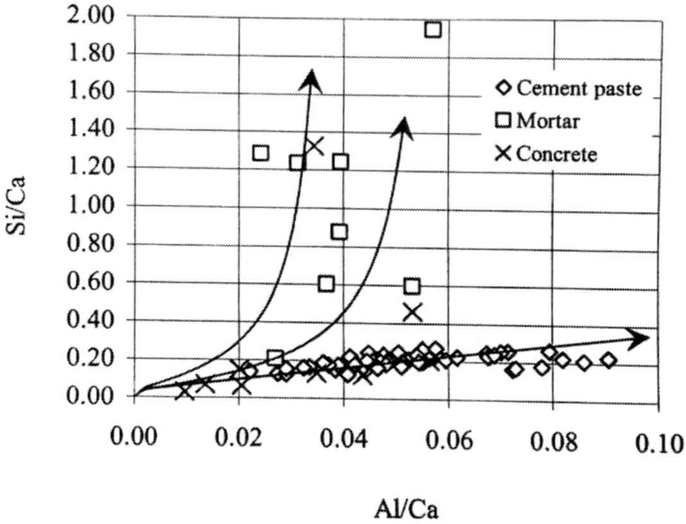

Figure 4.19. Atom ratio of Si/Ca versus Al/Ca of cement paste, mortar and concrete after applying microwave energy with different power levels.

The EDX results for the cement pastes subjected to microwave energy are shown in Figure 4.19: The atom ratio of Si/Ca versus Al/Ca for the pastes of (1CW/C_P0.38) after the application of microwave power of 100 watt for 30 min shows that for the normal paste, the Si/Ca cluster versus the Al/Ca cluster takes place in the narrow range of Si/Ca 0.163–0.240, but with a wider range of Al/Ca equal to 0.022–0.090. In the mortar and concrete, the degrees of scatter for Si/Ca versus Al/Ca are greater than those of the other pastes. It is likely that the aggregate of the paste affects the distribution of the Si, Ca, and Al elements, causing them to become more widely dispersed.

Figures 4.20 and 4.21 show the X-ray patterns of the hydrated products in the mortar and concrete after microwave power of 100 watt had been applied for 30 min. For mortar, the phases identified include hatrurite (Ca_3SiO_5) and lanite (Ca_2SiO_4). Similar to mortar, the concrete consists mainly of hatrurite (Ca_3SiO_5) and quartz (SiO_2).

The compressive strengths of the mortar of (3CAM0.38_1:2.75) when subjected to microwave energy, and concrete (3CAC0.38_1:1:1) compared to normal curing are shown in Figure 4.22: when cured at elevated temperatures using an microwave power level of 180 watt, mortar can develop strength quite rapidly, and with a higher microwave power level the development of strength decreases. It could be that an increase in microwave power significantly affects the interfacial transition zone between the neat paste and the aggregate. From the literature, increases in temperature expand the range of the interfacial transition zone and as a consequence compressive strength is reduced.

Similar to the mortar case, the microwave-cured concrete corresponds to the microwave power level shown in Figure 4.23, the compressive strength of concrete decreases with increases in the power level, as previously discussed earlier.

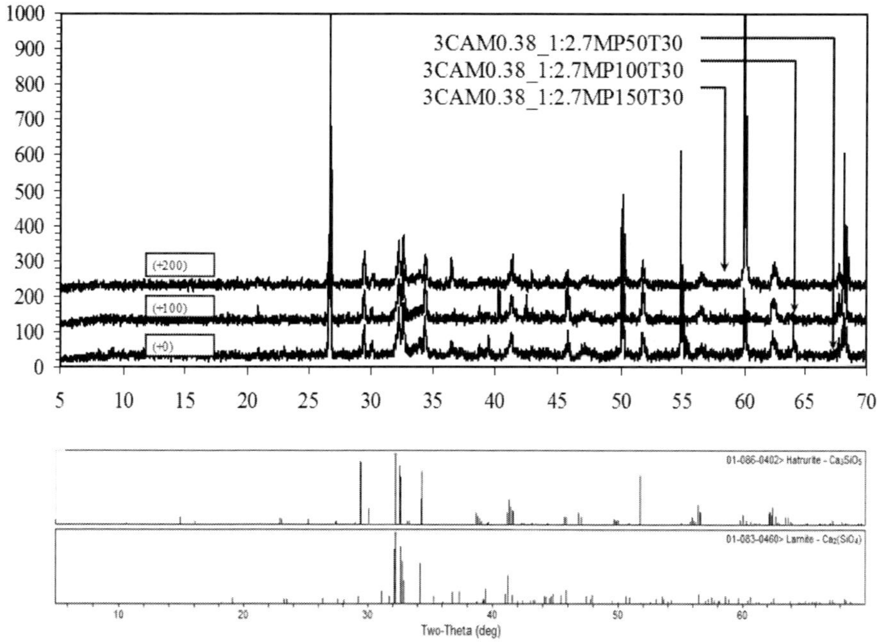

Figure 4.20. X-ray diffraction of mortar after applying microwave energy with different power levels.

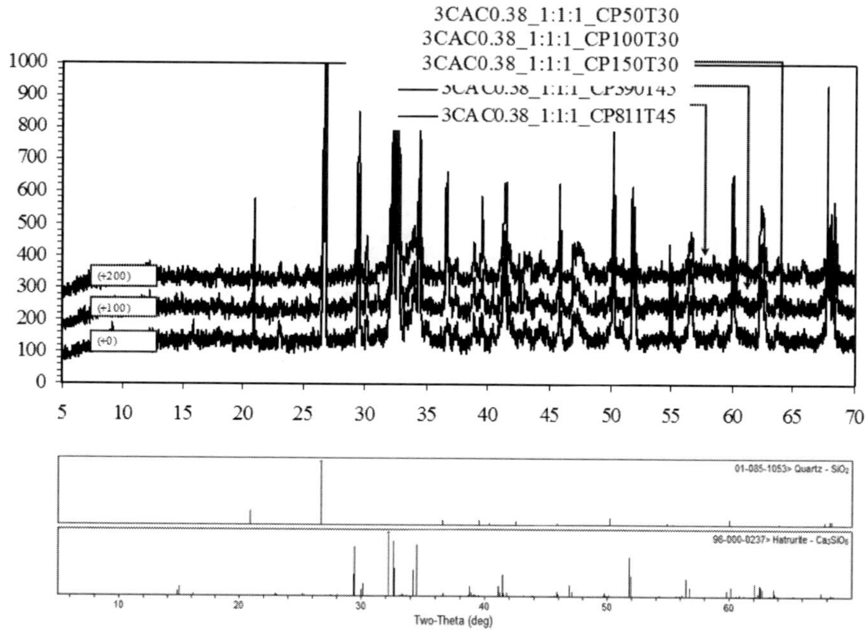

Figure 4.21. X-ray diffraction of concrete after applying microwave energy with different power levels.

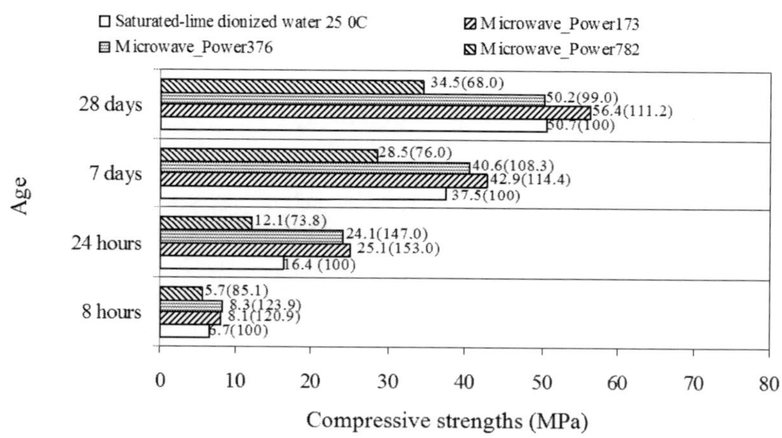

Figure 4.22. Compressive strengths of mortar after applying microwave energy with different power levels.

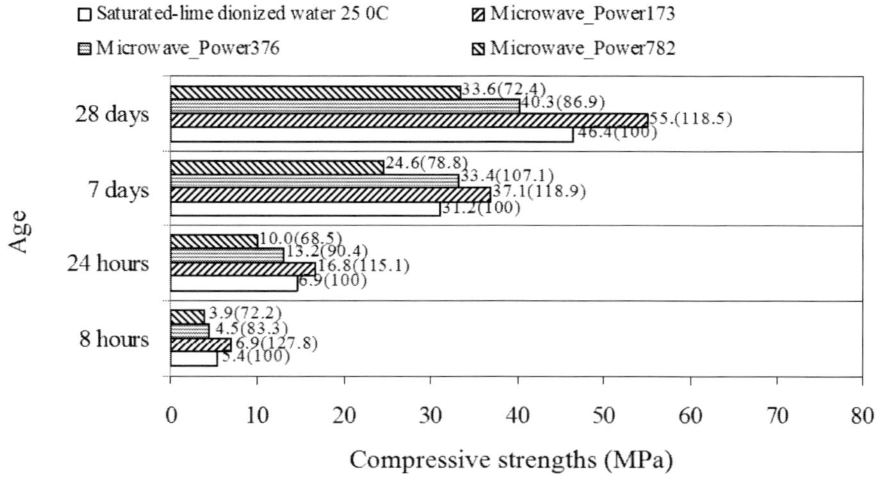

Figure 4.23. Compressive strengths of concrete after applying microwave energy with different power levels.

4.4.2. Effects of Microwave Processing

4.4.2.1. Delay Times

This section discusses the effects of delay times. Basically, the delay time is the time that elapses after mixing has finished and before the application of microwave energy has begun. The delay time comprises the dormant period, during which the hydration reaction slow down; the initial setting time, during which the mix begins to take on a form, and the final setting time, the point at which the structure has formed. Specific delay times are expected to affect various properties in particular ways. For example, if microwave were applied at the final setting time, pore content would be expected to increase so that microcracking would take place in the paste to be cured.

Figures 4.24–5.35 show the temperature profile against the application times for reinforced cement paste, FA–cement paste, MK–

cement paste, and SF–cement paste at water-cement ratio (w/c) of 0.38 with various delay times when subjected to a microwave power level of 100 watt for 30 min. In the case of the plain cement paste (Figure 4.24), microwave energy applied to the sample at a delay time of 30 min caused the highest temperature rise in the pastes to be processed—a result of the high water content especially the free water content within the sample. Because of the high water content, the sample at 30 min was able to absorb a large amount of energy and convert it easily into heat. However, applications at the initial setting time tended to decrease the extent to which the temperature rose due to the free water used in the hydration reaction and then combined in the C–S–H structure. However, even so, when microwave energy was applied at the final setting time the temperature increased to a greater extent than it had during the initial setting time. These relative increases can be accounted for by the fact that at the initial setting time point, temperature rise accrues from the initial state of the heat liberated from the hydration reaction, while at the final setting time temperature rise accrues from the approaching hydration peak, which sustains the heat generation from hydration. As a result, the temperature rise at the final setting time is higher than that of taking place at the previous initial setting time.

The cement pastes mixed with FA (Figure 4.25), MK (Figure 4.26), or SF (Figure 4.27) showed a similar tendency in regard to temperature rise as that of the reinforced cement paste. However, for all these mixes, the temperature rise at the final setting time was lower than that for the pastes at the initial setting time. The reason for this is that the materials used to replace the pozzolan material caused heat liberation to occur because of hydration is delayed more than the pozzolan used in the plain paste.

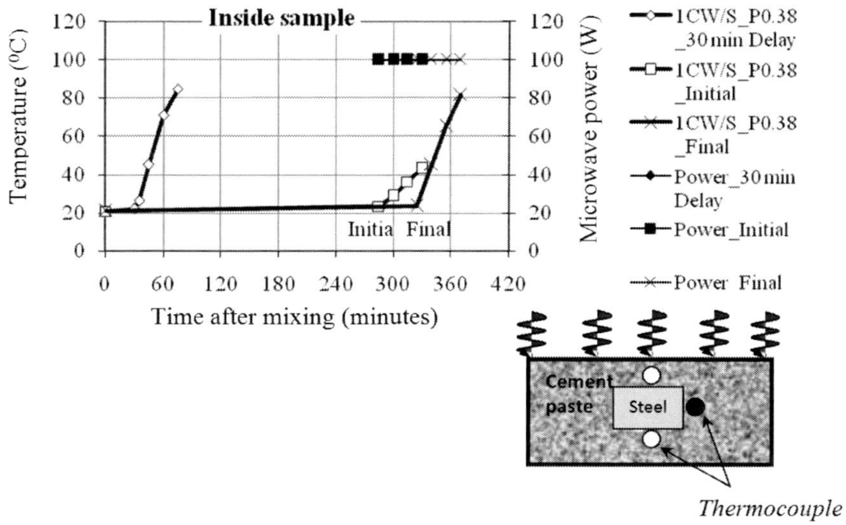

Figure 4.24. Temperature and power history during applying microwave energy of reinforced cement pastes at a water-to-solid material of 0.38 with vary delay times.

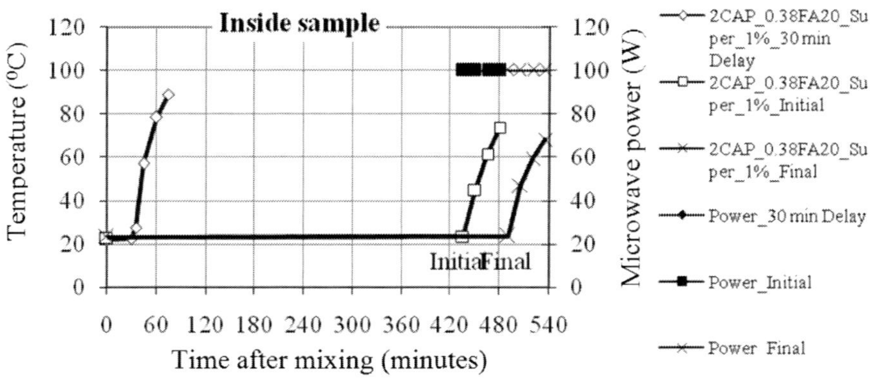

Figure 4.25. Temperature and power history during applying microwave energy of reinforced FA- cement paste at a water-to-solid material of 0.38 with vary delay times.

Typical micrographs of the plain cement paste (1CW/C_P0.38), FA–cement paste, MK–cement paste, and SF–cement paste when subjected to microwave energy at an microwave power level of 100 watt for 30 min are shown in Figure 4.28. Formed when microwave energy was applied at the final setting time, ettringite appears on the surface of the plain cement paste. On the other hand, the pozzolan paste showed no difference between the initial and final setting time of the microwave application.

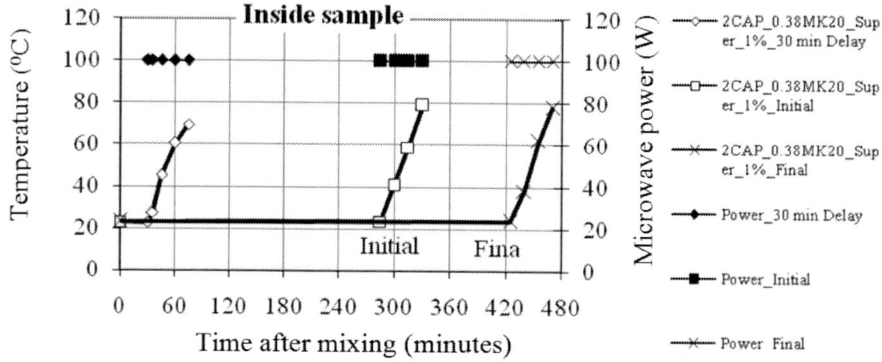

Figure 4.26. Temperature and power history during applying microwave energy of reinforced MK-cement pastes at a water-to-solid material of 0.38 with vary delay times.

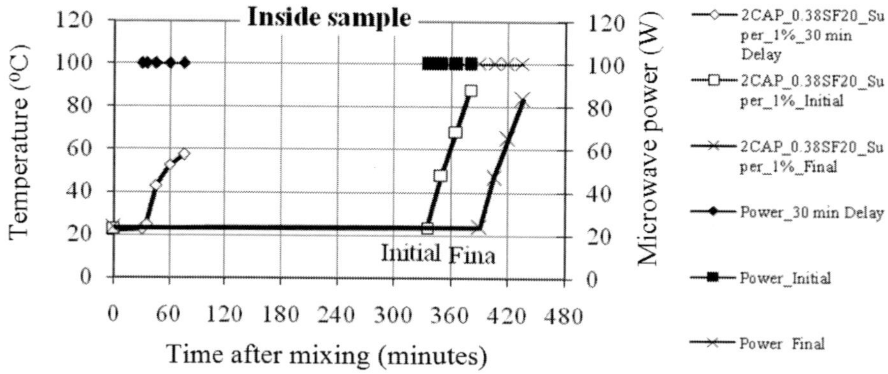

Figure 4.27. Temperature and power history during applying microwave energy of reinforced SF-cement pastes at a water-to-solid material of 0.38 with vary delay times.

From the EDX results for plain cement paste subjected to microwave energy are shown in Figure 4.29. The atom ratio of Si/Ca versus Al/Ca for the pastes of (1CW/C_P0.38) to which microwave power of 100 watt was applied for 30 min, agglomerated and tended to show an increase in the Al/Ca ratio, indicating that microwave energy can cause Al-element distribution. For the FA–cement paste, the distribution of Al/Ca ratio showed a narrower range than that did that of the plain paste. In the MK–cement paste, the distribution was wider than it was in the plain paste because of a major constituent of MK is aluminum. While the SF in SF–paste consists mainly of Si-element.

Figures 4.30 – 4.33 show the effects of delays on the phase characteristics of the 1CW/C_P0.38, CAP_0.38FA20_Super_1%, 2CAP_0.38MK20_Super_1%, and 2CAP_0.38SF20_Super_1% after microwave power of 100 watt had been applied for 30 min. The calcium silicate hydrate (Ca_3SiO_5), calcium hydroxide ($Ca(OH)_2$) lime (CaO), and Xenotile phases were found to be similar.

(a) 1CW/C_P0.38 at microwave power 100 watt for 30 min and delay time at initial setting time.

(b) 2CAP_0.38FA20_Super_1% microwave power 100 watt microwave power 100 watt for 30 min and delay time at initial setting time.

(c) CAP_0.38MK20_Super_1% microwave power 100 watt for 30 min and delay time at initial setting time.

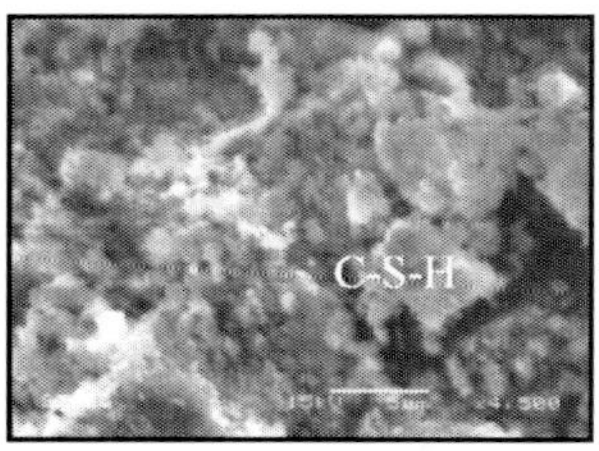

(d) 2CAP_0.38SF20_Super_1% at microwave power 100 watt for 30 min and delay time at initial setting time.

Figure 4.28. (Continued).

(e) 1CW/C_P0.38 at microwave power 100 watt for 30 min and delay time at final setting time.

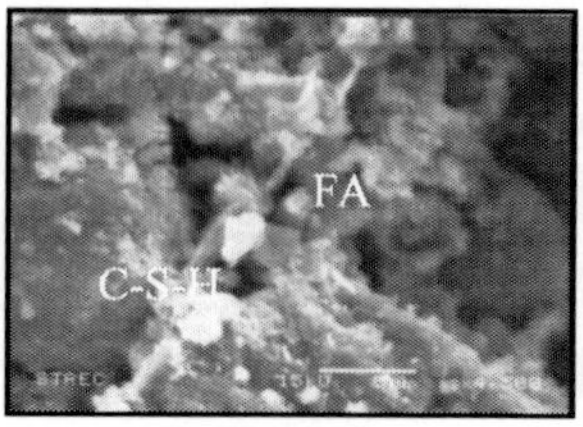

(f) 2CAP_0.38FA20_Super_1% at microwave power 100 watt for 30 min and delay time at final setting time.

(g) 2CAP_0.38MK20_Super_1%at microwave power 100 watt for 30 min and delay time at final setting time.

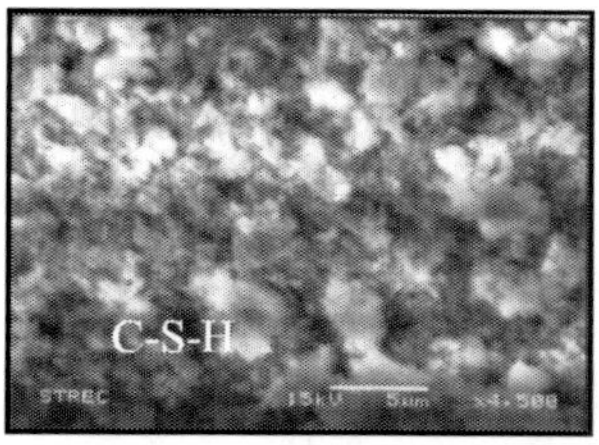

(h) 2CAP_0.38SF20_Super_1% at microwave power 100 watt for 30 min and delay time at final setting time.

Figure 4.28. Micrographs of various cement pastes at different delay times.

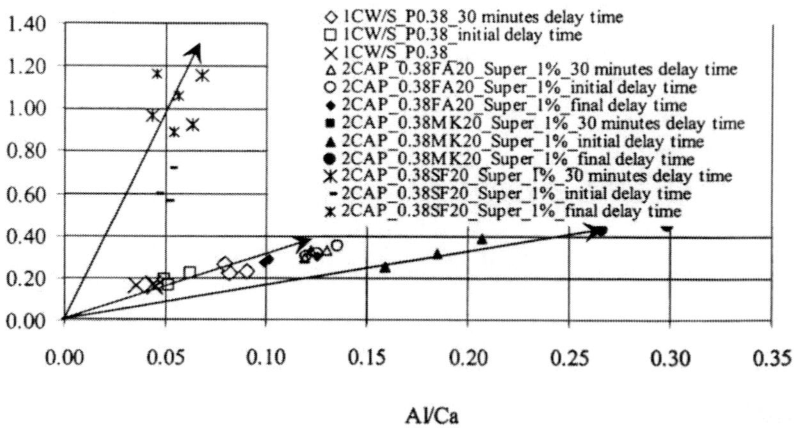

Figure 4.29. Atom ratio of Si/Ca versus Al/Ca of various cement pastes with different delay times when applying microwave energy.

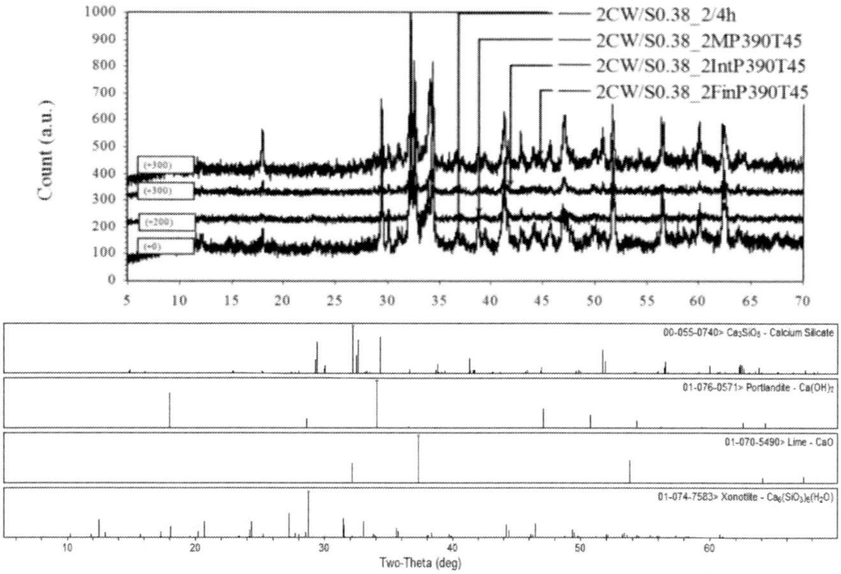

Figure 4.30. X-ray diffraction of 1CW/C_P0.38 cement paste with different delay times when applying microwave energy.

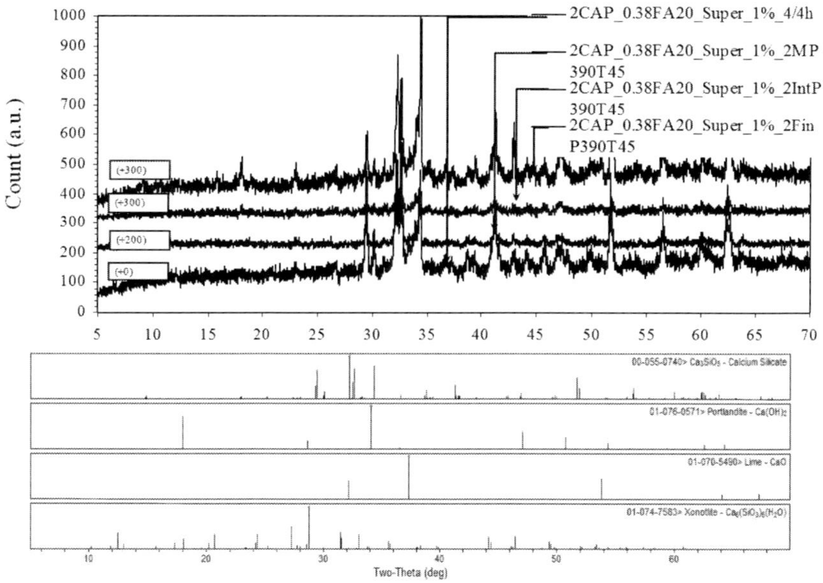

Figure 4.31. X-ray diffraction of 2CAP_0.38FA20 reinforced cement pastes with different delay times when applying microwave energy.

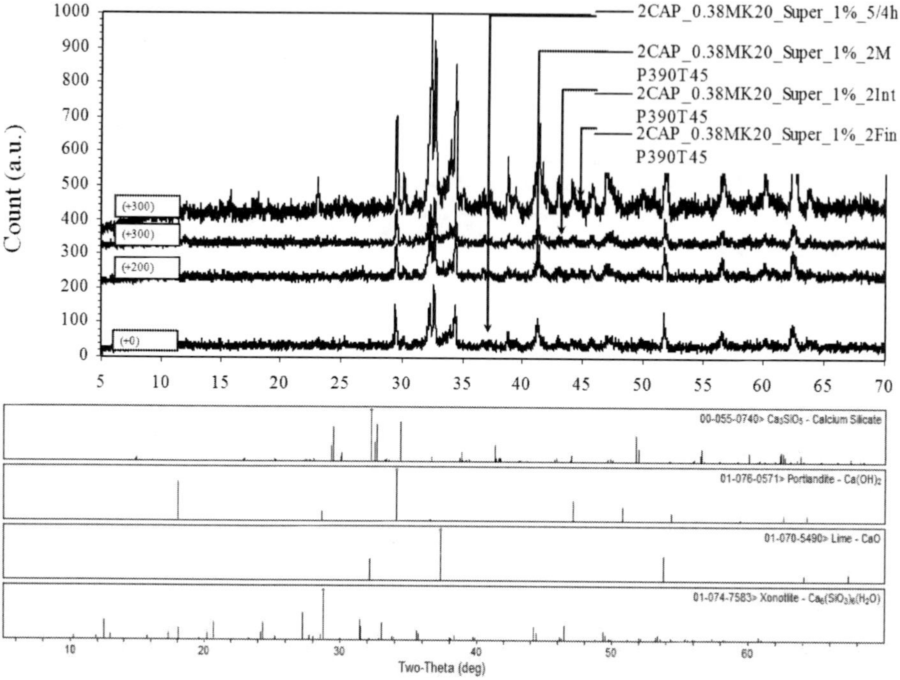

Figure 4.32. X-ray diffraction of 2CAP_0.38MK20 reinforced cement pastes with different delay times when applying microwave energy.

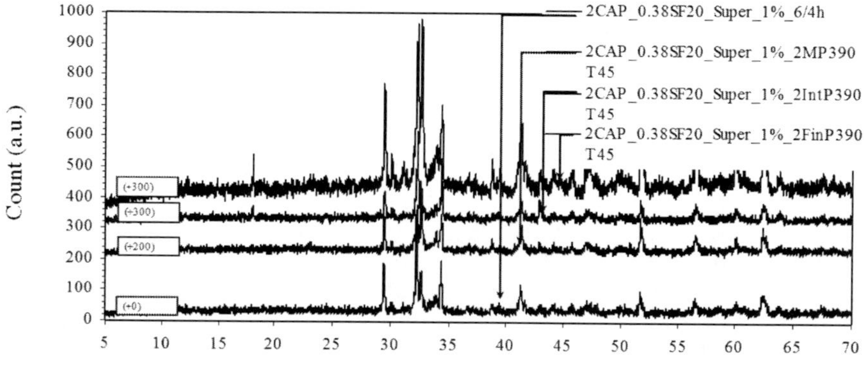

Figure 4.33. (Continued).

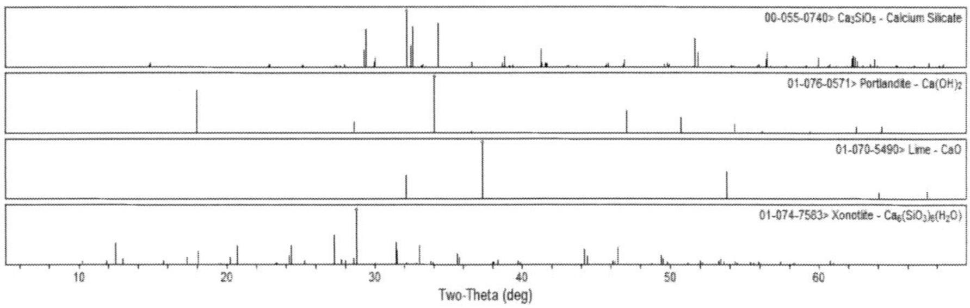

Figure 4.33. X-ray diffraction of 2CAP_0.38SF20 reinforced cement pastes with different delay times when applying microwave energy.

The compressive strengths of the 1CW/C_P0.38, CAP_0.38FA20, 2CAP_0.38MK20, and 2CAP_0.38SF20 after microwave power of 100 watt had been applied for 30 min are shown in Figure 4.34. Overall, it can be concluded that using microwave energy for curing beginning at the final setting time can reduce compressive strengths of both early- and late-age reinforced and pozzolan pastes. This is because heat generated by microwave imposed after the C–S–H structure has formed increases the temperature difference between the microstructured paste and the paste during formation, and as a result cracking takes place immediately. Moreover, the remaining free water in the capillary pores shows a higher temperature increase than do the hydration products which cause cracking in the internal capillary pores that penetrate the C–S–H structure deeply. Consequently, these effects cause a reduction in the compressive strength of all the pastes.

Figure 4.35 presents the results of the remaining $Ca(OH)_2$ content of the reinforced cement pastes (1CW/C_P0.38), which are 2.07% (w/w) and 2.34% (w/w) for the initial and final setting times, respectively. These results indicate that applying microwave energy at the final setting time can increase the amount of $Ca(OH)_2$ more so than applying it at the initial setting time. It could be that during the final setting time, the heat generated by microwave in conjunction with the heat of hydration causes an even greater amount of $Ca(OH)_2$.

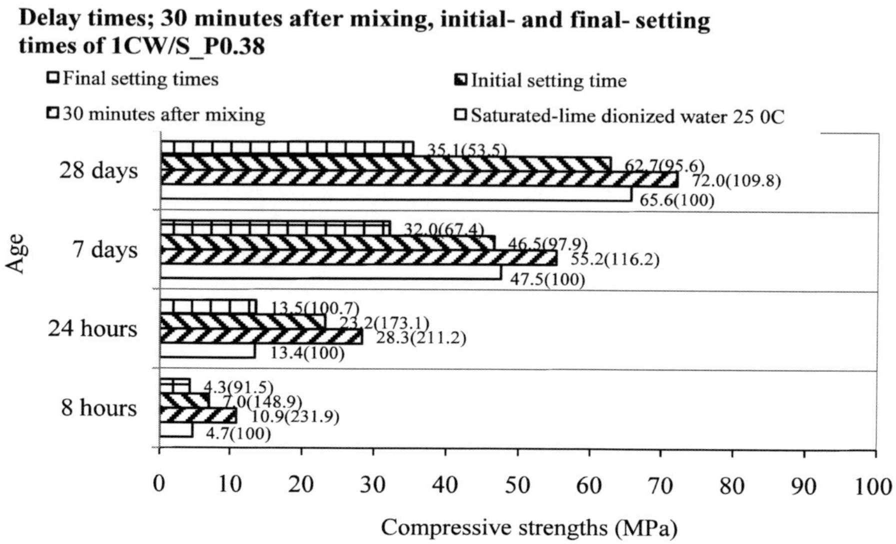

(a) Reinforced pastes.

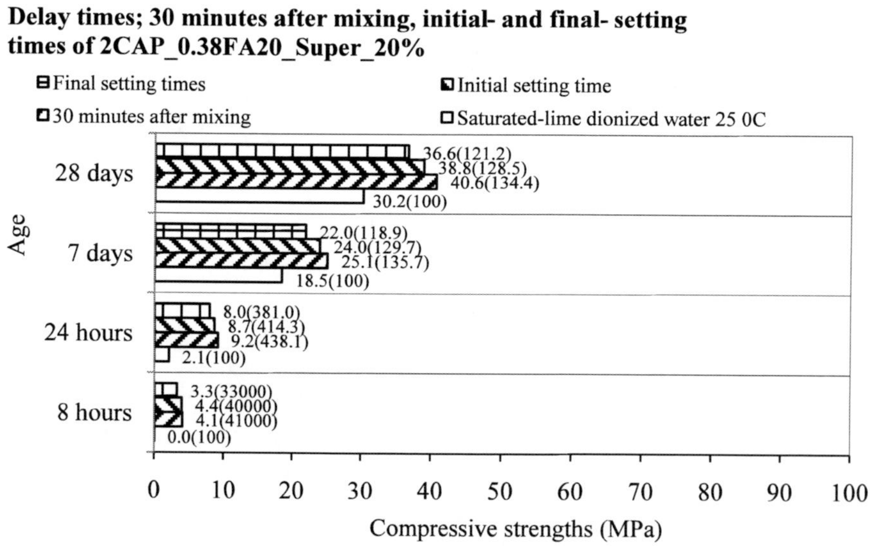

(b) Reinforced pastes mixed with FA.

Figure 4.34. (Continued).

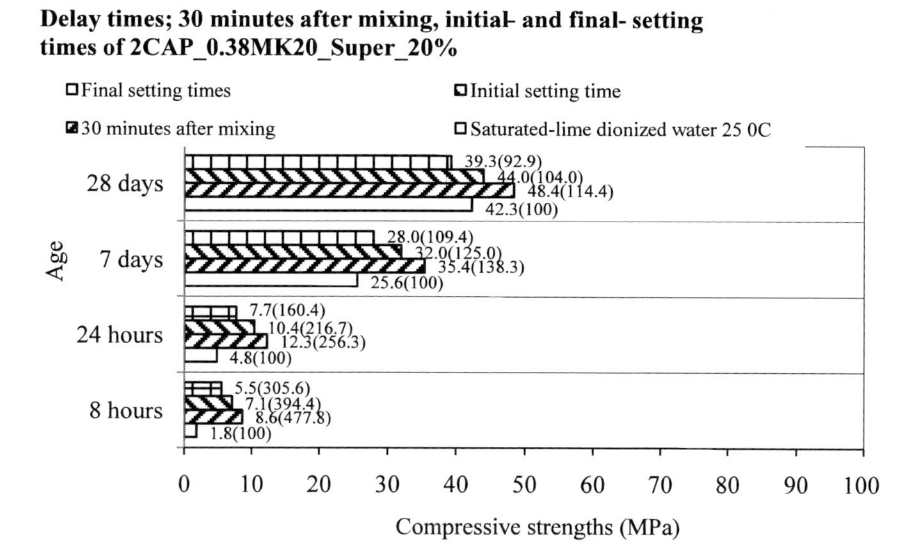

(c) Reinforced pastes mixed with MK.

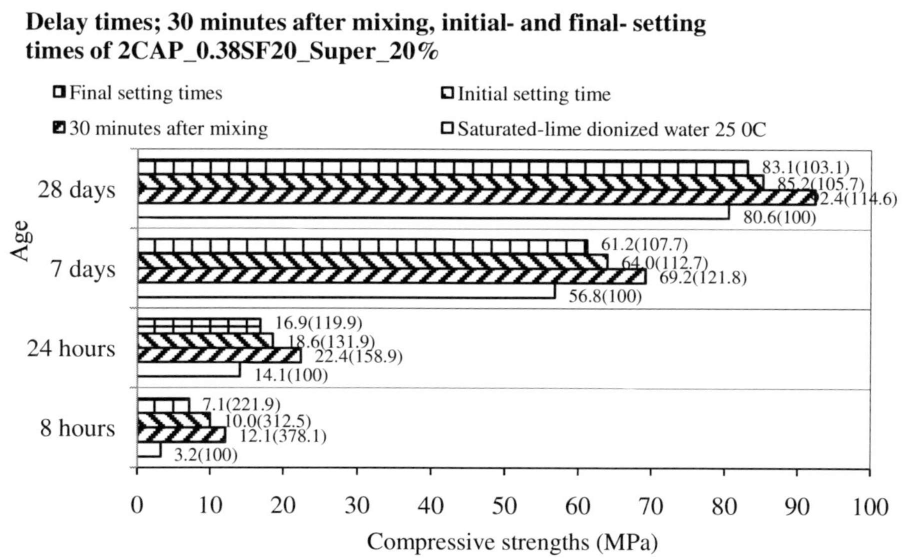

(d) Reinforced Pastes mixed with SF.

Figure 4.34. Compressive strengths of 1CW/C_P0.38 reinforced cement paste with different delay times when applying microwave energy.

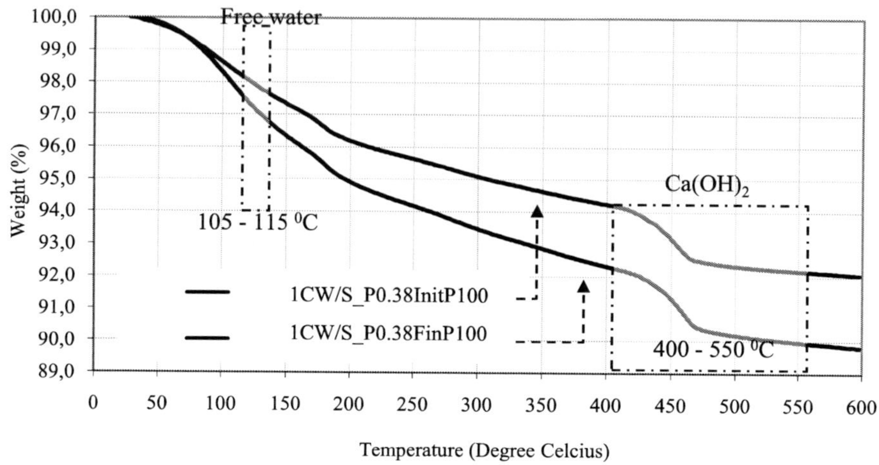

Figure 4.35. Thermogravimetric analysis (TG) results by comparison among different delay times.

4.4.2.2. Power Levels and Application Times

This section discusses the effects of different microwave power levels and application times. It is well-known that the level of the microwave power applied affects the generation of heat, and that even at high levels of microwave power the duration of the application must also be sufficient to effect curing.

Figure 4.36 shows the temperature inside the microwave-cured reinforced paste samples and their power history in regard to the applying microwave energy at 50 watt for 30 min, 150 watt for 15 min, and 50 watt for 90 min. It was also found that the rate at which the temperature increases is related to a decreasing water-cement ratio; that is, the sample with the low water-cement ratio increases its temperature more rapidly than do the other samples. For the same reason aforementioned, the sample with the lowest water-to-cement ratio has higher concentrations of Ca^{2+}, Si^{4+}, and so on, than do the samples with higher water-to-cement ratios.

When comparisons the rate of temperature rise at the early period on microwave curing, a higher microwave power can generate heat (temperature) inside the sample higher than that of the lower microwave power. It corresponds to high microwave power; that is high electric field intensity.

However, microwave energy applied for longer duration can induce a higher temperature rise than energy applied for a shorter duration. For example, microwave energy of 50 watt applied for 90 min shows a final temperature increase of 112°C and microwave energy of 150 watt applied for 15 min shows a final temperature increase of 62°C. This suggests that as the duration of microwave energy application increases so the heat accumulated by the microwave energy accumulates and is added to the heat produced by the hydration reaction. Further, microwave energy applied for a long enough period encompasses the initial and final setting times, which both accrue more heat hydration, the latter accruing the most, when compared to the dormant period.

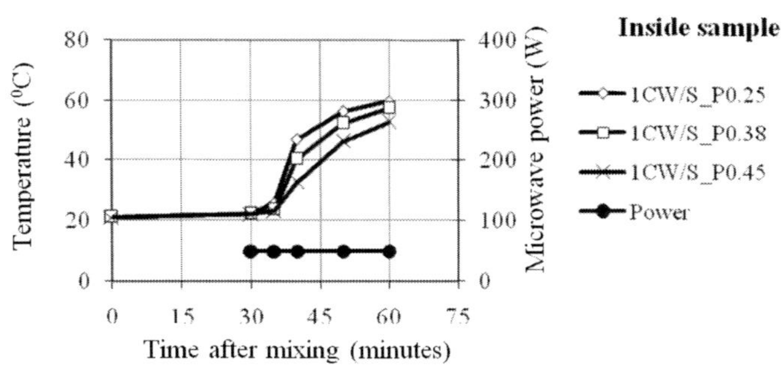

(a) Power level at 50 watt for 30 min.

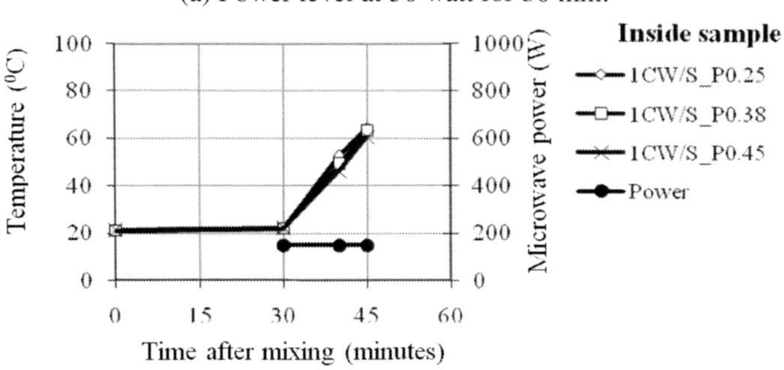

(b) Power level at 150 watt for 15 min.

Figure 4.36. (Continued).

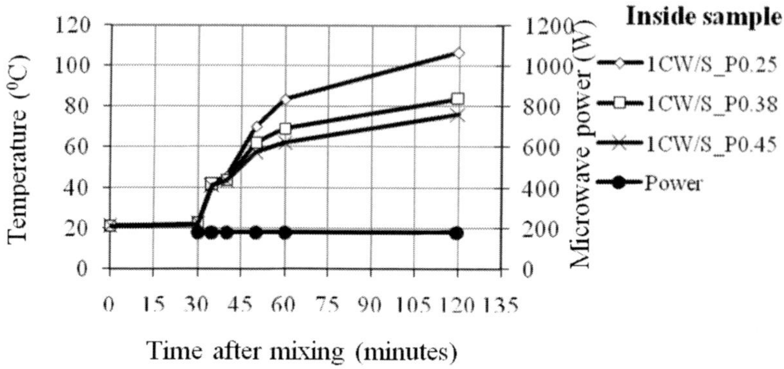

(c) Power level at 50 watt for 90 min.

Figure 4.36. Temperature and power history during applying microwave energy of various reinforced cement pastes with different power levels and times of application.

(a) 1CW/C_P0.38 after applying microwave energy power 100 watt for 30 min.

(b) 1CW/C_P0.38 after applying microwave energy power 150 watt for 15 min.

Figure 4.37. Micrographs of 1CW/C_P0.38 reinforced cement pastes with different power levels and application times of microwave energy.

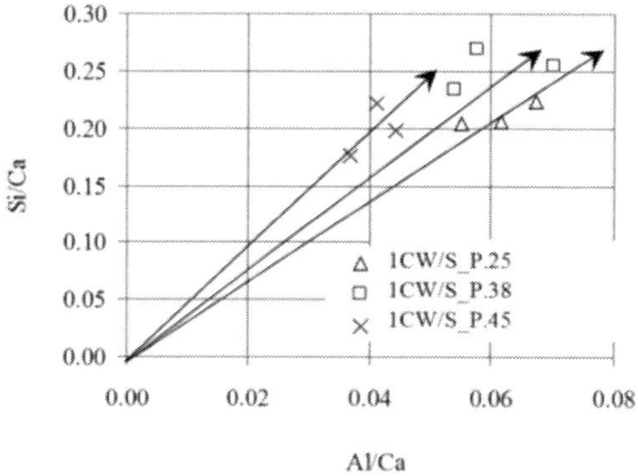

(a) Microwave power 50 watt for 30 min.

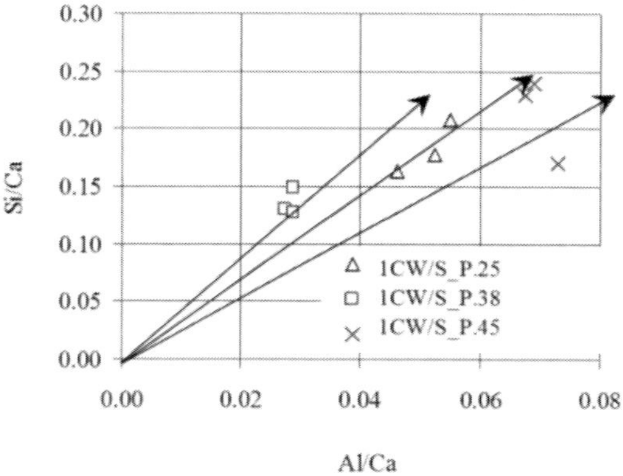

(b) Microwave power 50 watt for 90 min.

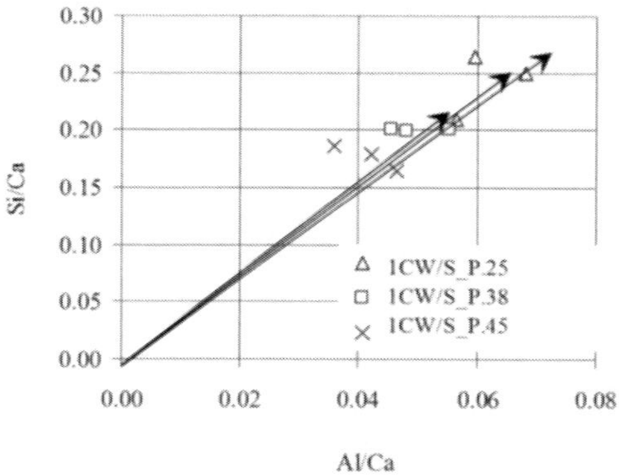

(c) Microwave power 150 watt for 15 min.

Figure 4.38. Atom ratio of Si/Ca versus Al/Ca of the pastes of 1CW/C_P0.25, 1CW/C_P0.38 and 1CW/C_P0.45 with different power levels and application times of microwave energy.

As shown in Figure 4.37, a comparison of the morphology of the paste 1CW/C_P0.38 subjected to microwave curing of 100 watt for 30 min with that subjected to 150 watt for 15 min shows that overall they are similar in shape. However, the paste subjected to microwave power of 100 watt for 30 min shows a more plate-like shape than does the paste subjected to 150 watt for 15 min. It may be that a higher microwave power causes evaporation to take place immediately and then effects agglomeration in the paste sample. In other words, the rapid drying of the water present in the sample may cause the cement particles covered by the water molecules to bind together.

Figure 4.38 shows the atom ratio of Si/Ca versus Al/Ca for the 1CW/C_P0.25, 1CW/C_P0.38, and 1CW/C_P0.45 pastes with different power levels and application times of microwave energy. In 0.25 w/c cement paste, an increase in the microwave power level does not affect the Si/Ca and Al/Ca ratios; however, an increase in the power level with a longer time of application decreases the Si/Ca and Al/Ca ratios. It can clearly be seen that both Si and Al elements are sensitive to electromagnetic waves. And, the same 0.38 w/c cement paste manifests the same behavior as does the 0.25 w/c cement paste. Nevertheless, the 0.45-w/c cement paste exhibits the opposite feature; that is, a longer application times increases the Si/Ca and Al/Ca ratios. This means that the presence of a comparatively large quantity of water, especially capillary water, in the cement-paste system affects the formation and configuration of the paste's products (C–S–H).

The effect of different power levels and application times of microwave energy on the phase characteristics of 1CW/C_P0.38 is shown in Figure 4.39. It is found that the calcium silicate hydrate (Ca_3SiO_5), calcium hydroxide ($Ca(OH)_2$) lime (CaO) and Xenotile phases are similar.

The compressive strengths of 1CW/C_P0.25, 1CW/C_P0.38, and 1CW/C_P0.45 with a power level of 173 watt for 45 min, 173 watt for 90 min, and 782 watt for 15 min are shown in Figure 5.48. The pastes show similar behavior in regard to compressive strength development; that is, compressive strength develops most quickly in the first 28 days.

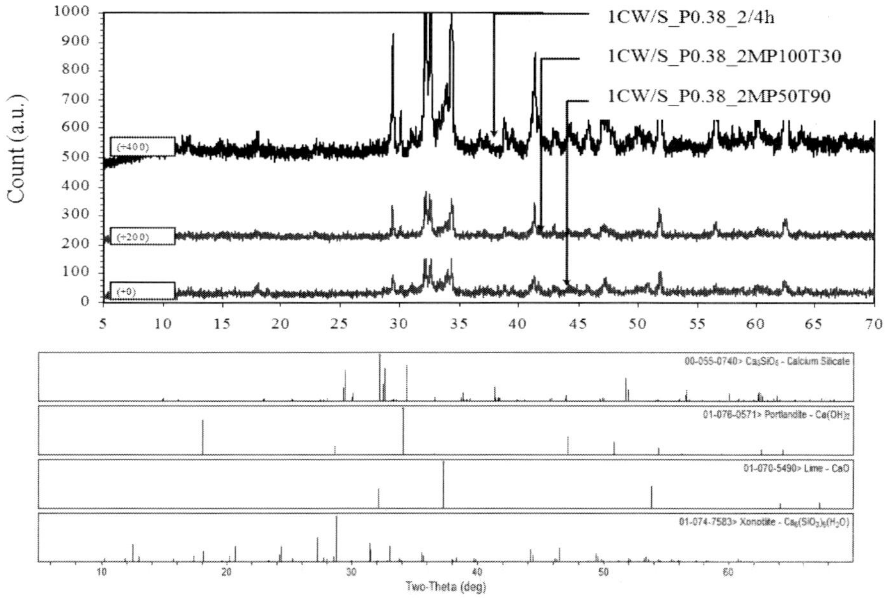

Figure 4.39. X-ray diffraction the pastes of 1CW/C_P0.38 with different power levels and application times of microwave energy.

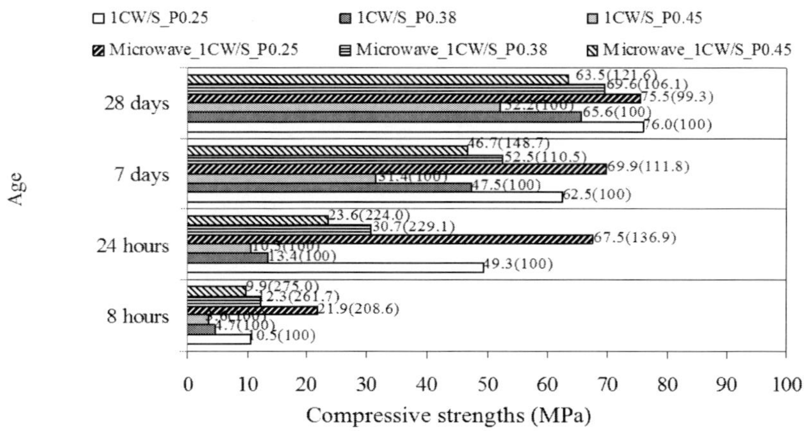

(a) Reinforced pastes when subjected to a power level 50 watt for 30 min.

Figure 4.40. (Continued).

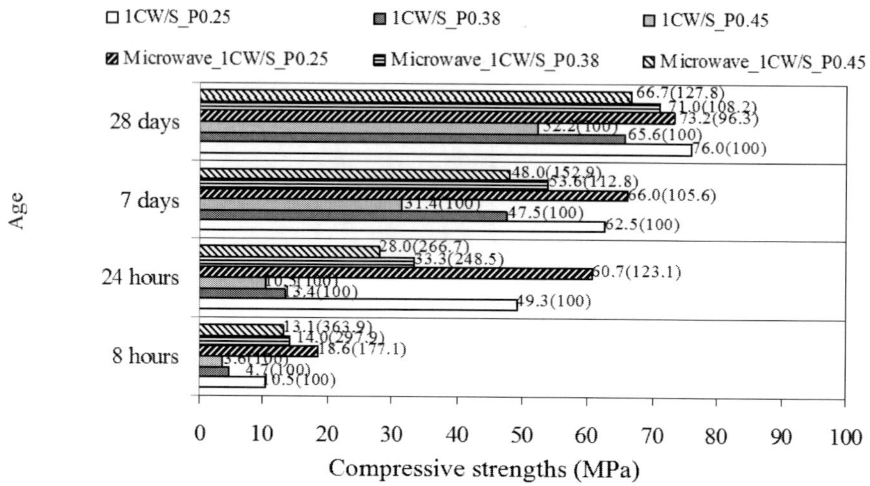

(b) Reinforced pastes when subjected to a power level 50 watt for 90 min.

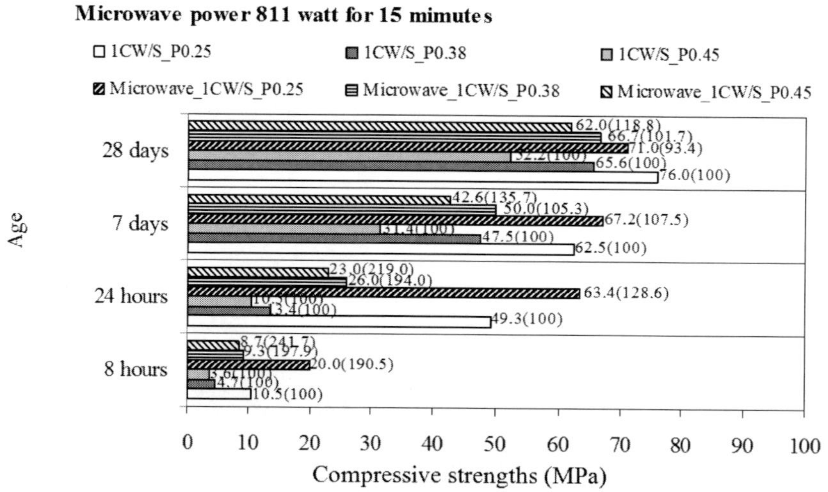

(c) Reinforced pastes when subjected to a power level 150 watt for 15 min.

Figure 4.40. Compressive strengths the reinforced pastes of 1CW/C_P0.25, 1CW/C_P0.38 and 1CW/C_P0.45 with different power levels and application times of microwave energy.

The different power levels and application times of microwave energy affect the compressive strength of the microwave-cured pastes. A higher microwave power level causes C–S–H to take on an irregular shape, and it also causes thermal cracking (Haddad and Al-Qadi, 1998).

It can be seen that at 28-days, the compressive strength of the 0.25 w/c microwave-cured paste shows a lower value than do the other samples. However, when the water-cement ratio of the paste increases, this effect is lessened.

4.5. Conclusion

Microwave energy can accelerate hydration reaction or increase the temperature within the microwave-cured reinforced cement sample; the effects of water-to-solid mass ratios, pozzolan materials, reinforced mortar and concrete, delay times, power level, and application duration were taken into account. The test results showed that the temperature increases monotonically among the positions of measurement during the microwave-curing process. The typical micrographs of the microwave-cured reinforced paste at the age of 4 h after mixing, 28 days after curing in lime-saturated deionized water, and when subjected to microwave energy showed that the samples consisted of hydrated phases and pores, as well as cores of $Ca(OH)_2$ dendrite crystals or other crystals (marked CH), C–S–H, and granular structure. Furthermore, some ettringite (Aft) was found in the samples cured by microwave energy. From the SEM–EDS results it was observed that although the measured Ca/Si ratios of the pastes were similar in magnitude, they consistently decreased when the temperature decreased. For compressive strengths, the microwave-cured reinforced pastes developed strength quite rapidly, which correspond to the maintenance of the microwave power and time of application, and the water content of the pastes' hydration products were affected depending on the w/c; in addition, the $Ca(OH)_2$ contents were affected due to the completion of hydration reaction.

Chapter 5

A Theoretical Model to Predict the Properties of the Reinforced Cement-Based Materials when Subjected to Microwave Energy

5.1. Electromagnetic Theory

The electromagnetic (microwave) fields include electric and magnetic fields that move simultaneously with each other through space. This movement can be represented simply by plane waves, as in Figure 5.1. The relationship between time-varying electric and magnetic fields can be described macroscopically by using Maxwell's equations. The important observations are (a) the displacement current density is simply the rate at which the electric flux density $\vec{D}$ varies with time, (b) a time-varying electric field creates a time-varying magnetic field due to the action of $(\partial\vec{D}/\partial t)$ as a source of the magnetic field, and (c) the time-varying magnetic and electric fields are interdependent. With the help of boundary characteristics, the propagation conditions of plane waves are assumed to be as follows: (a) the component of the field quantities $\vec{E}$ and $\vec{H}$ lying in the transverse plane, a plane perpendicular to the direction of the propagation of the wave; and (b) the amplitude attenuation of the wave is reduced by the familiar factor of e^{-1} (0.368) in a dielectric medium (a type of material, such as concrete, brick, or paper, that can absorb microwave energy and then can be converted partially into other form of energy) by means of creating ion oscillations and continuing dipole relaxation.

The electromagnetic (microwave) waves propagate in free space at the speed of light ($c = 2.9979x10^8\ m/s$). Further information on the form of these waves can be obtained by noting that the $\vec{E}$ - and $\vec{B}$ - fields must also satisfy Maxwell's equations. To consider a specific type of wave, although the results derived are general: This wave is said to be an unbounded plane indicating that there are existing planes (or wave fronts) perpendicular to the direction of propagation, over which all quantities of the wave are constant, as shown in the plain wave in Figure 5.1.

One of the major achievements of Maxwell's equations is their ability to correctly predict electromagnetic waves from considerations of the compacted forms of the four equations. The equations' validity is based on their consistency with all the existing experiments concerning electromagnetic phenomena. The physical meanings of these equations are listed identically in identical differential forms below:

(a) Gauss's law is a mathematical expression of the experimental fact that electric charges repel or attract one another with a force proportionally inverse to the square of the right distance between them (Eq.(5.1)):

$$\nabla.\vec{D} = q \tag{5.1}$$

(b) Faraday's law is based on time-varying the magnetic flux inducement of electromotive force (Eq.(5.2)):

$$\nabla x\vec{E} = -\frac{\partial \vec{B}}{\partial t} \tag{5.2}$$

(c) Ampère's law states that the line integral of the magnetic field over any close contour must equal the total current enclosed by that contour (Eq. (5.3))

$$\nabla x\vec{H} = \vec{J} + \frac{\partial \vec{D}}{\partial t} \tag{5.3}$$

(d) The last of Maxwell's four equations relies on the fact that there are no magnetic charges or magnetic monopoles and that magnetic field lines always close on themselves (Eq.(5.4)):

$$\nabla.\vec{B} = 0 \tag{5.4}$$

where $\vec{E}$ is the electric field intensity (V/m), $\vec{H}$ is the magnetic field intensity (A/m), $\vec{D}$ is the electric flux density (C/m), $\vec{B}$ is the magnetic flux density (Wb/m^2), t is the time, and q is the electric charge density (C/m^3).

The constitutive relations that relate $\vec{H}$ and $\vec{B}$, $\vec{E}$ and $\vec{D}$, and $\vec{E}$ and $\vec{J}$ are shown in Eqs.(5.5) – (5.7) respectively. Then, substituting (Eq.(5.5) – (5.7)) into (Eqs.(5.1) – (5.4)) results in (Eqs.(5.8) – (5.10):

$$\vec{B} = \mu\vec{H} \tag{5.5}$$

$$\vec{D} = \varepsilon\vec{E} \tag{5.6}$$

$$\vec{J} = \sigma\vec{E} \tag{5.7}$$

where μ is the magnetic permeability (H/m), σ is the electric conductivity (1/Ohm), and ε is the dielectric constant (F/m).

$$\nabla.\vec{E} = \frac{q}{\varepsilon} \tag{5.8}$$

$$\nabla x\vec{E} = -\mu\frac{\partial\vec{H}}{\partial t} \tag{5.9}$$

$$\nabla x\vec{H}=\sigma\vec{E} + \varepsilon\frac{\partial\vec{E}}{\partial t} \tag{5.10}$$

$$\nabla.\vec{H} = 0 \tag{1.12}$$

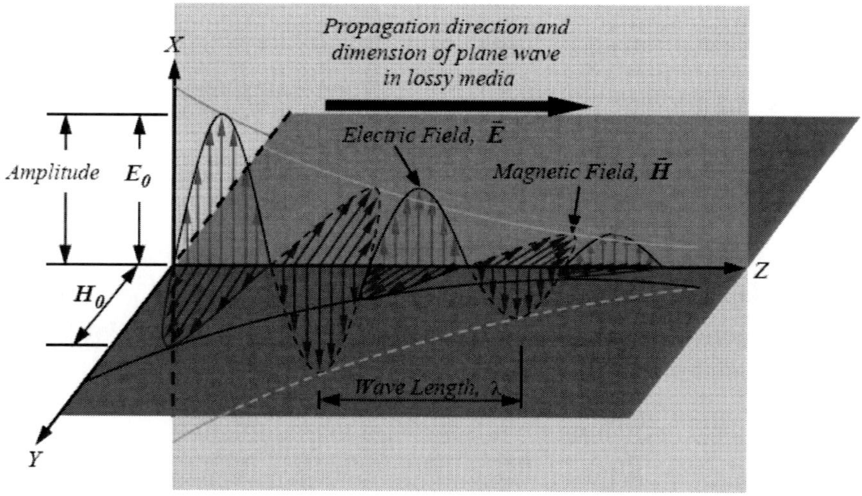

Figure 5.1. Electromagnetic wave propagation.

5.2. NUMERICAL ANALYSIS

Our research concept is to use microwave energy to improve the quality of concrete. Microwave energy generates heat within a material through rapid changes in polarization as indicated in Equation (5.11).

$$Q=\sigma\left|\vec{E}\right|^2=2\pi f\varepsilon_0\varepsilon_r^{'}\left(tan\sigma\right)E^2 \tag{5.11}$$

where Q is the microwave energy, σ is the effective conductivity, f is the frequency (Hz), ε_0 is the permittivity of free space (8.8514×10^{-12} farads/meter), $\varepsilon_r^{'}$ is the relative dielectric constant, $tan\ \delta$ is the loss tangent coefficient, and $\vec{E}$ is the electric field intensity (volts/meter).

Equation (5.11) describes the relationship between the intensity of the electric field, the microwave frequency, and the dielectric properties of the material and their effect on volumetric heat generation. Another factor is the penetration depth (*Dp*), which is the depth to which the radiation penetrates the material and still retains 37% (*1/e*) of the energy at the surface, calculated using Equation (5.12) and figure 5.2 (Rattanadecho). If the penetration depth is less than the thickness of the material, resonance phenomena occur and the electric field distribution does not conform to Lambert's law, resulting in local increases in the intensity of the electromagnetic field.

Equation (5.12) is applicable to materials that are sufficiently thick to be considered infinite. Microwaves rapidly generate heat through the large fields developed in a standing wave created by the forward propagating waves and the waves reflected from the air-solid interface. When such a phenomenon occurs, it cannot use these rules to predict wave reflection, so the variation in small pieces to use Maxwell's equations into the analysis. The distribution of temperature, velocity and absorbability of microwave energy are described. It's however not considered to be porous material.

$$D_p = \frac{1}{\frac{2\pi f}{\upsilon}\sqrt{\frac{\varepsilon_r'\left(\sqrt{1+(\tan\delta)^2}-1\right)}{2}}} \tag{5.12}$$

where D_p is the penetration depth, ε_r'' is the relative dielectric loss factor, and υ is the microwave propagation velocity in the dielectric material determined from the relation $c/\sqrt{\varepsilon_r'}$.

5.2.1. Analysis of Dielectric Properties

The dielectric properties can be represented in Table 5.1.

Table 5.1. Dielectric permittivity of 1CW/C_0.38 paste

Water-cement ratio (w/c) = 0.38		
Time after mixing (at h)	ε'	ε''
0	15.1288	4.8219
0.5	**15.0861**	**4.4458**
1	14.8792	4.2154
3	14.5707	4.1633
6	12.4870	3.2410
9	5.6590	0.9733
12	4.4850	0.4793
15	3.7910	0.5140
18	3.1623	0.3777
21	4.0093	0.5853
24	4.2740	0.7100

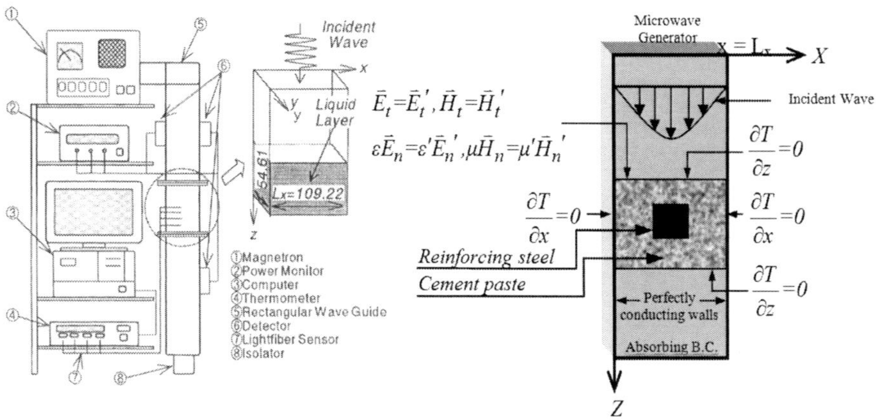

Figure 5.2. The distribution of microwave radiation in the XZ plane.

5.2.2. Analysis of Heating Problem: Conduction Mode of Heat Transfer

A schematic diagram of the physical model is shown in Figure 5.2. The temperature of the material introduced to the incident wave is obtained by solving the conventional heat transport equation with the microwave power included as a local electromagnetic heat-generation term.

5.2.1.1. Assumptions

The materials considered were cement pastes at a w/c equal to 0.38, which consist of Portland cement Type I, water, and air. They are, therefore, homogeneous and isotropic in terms of structure. In order to analyze the process of heat transport due to microwave heating of dielectric materials, the following assumptions were made:

- The local thermodynamic equilibrium is achieved.
- The liquid phase is not compressible $\rho_l = constant$.

- No chemical reactions within the material are taken into account.
- Radiation modes are negligible.
- The effect of the natural and induced convections can be neglected.
- In this task, from a macro-level point of view, the pore structure within the microwave-cured paste is assumed to be homogeneous and isotropic. Therefore, a heating model for a homogeneous and isotropic is used in this analysis.

5.2.1.2. Basic Equations

The governing equations describing the temperature rise in the paste to be processed by microwave energy are:

Electromagnetic wave:

$$\nabla \times \left(\mu_r^{-1} \nabla \times \vec{E}\right) - k_0^2 \left(\varepsilon_r - j\frac{\sigma}{\omega\varepsilon_0}\right)\vec{E} = 0, \varepsilon_r = n^2 \tag{5.13}$$

Heat transfer (conduction mode):

$$\rho C_p \frac{\partial T}{\partial t} + \nabla \times \left(-k\nabla T\right) = Q \tag{5.14}$$

where μ_r is the relative permeability, $\vec{E}$ is the electric field intensity (volts per meter (V m^{-1})), k_0 is the free space wave number, ε_r is the relative permittivity, σ is the electric conductivity (siemens per meter (S·m^{-1})), ω is the angular frequency (radians per second), and n^2 is the relative permittivity.

k is the thermal conductivity (W/(m·K)), T is the absolute temperature (K), Q contains heat sources other than viscous heating

(W/m^3), ρ is the density (kg/m^3), and Cp is the specific heat capacity at constant pressure (J/(kg·K)).

5.2.1.3. Boundary Conditions

(a) Adiabatic boundary condition

Assuming that the surroundings of the paste are insulated (no heat and moisture transfer between the system and surroundings):

$$\left.\frac{\partial T}{\partial n}\right|_{sample\,surface} = \left.\frac{\partial T}{\partial t}\right|_{sample\,surface} = 0 \tag{5.15}$$

(b) Continuity boundary condition

For the microwave heating of the cement paste, the temperature and heat flux at the interface of sample and air within the microwave cavity to be continuity:

$$T = T', \quad \lambda_{eff}\left.\frac{\partial T}{\partial n}\right|_{boundary} = \lambda'_{eff}\left.\frac{\partial T}{\partial n}\right|_{boundary} \tag{5.16}$$

(c) Electromagnetic boundary condition

$$n \times \vec{E} = 0 \tag{5.17}$$

where n is the normal vector.

(d) The initial condition of the microwave-cured cement reinforced paste sample defined as:

$$T = T_0 = 298.15\,K \quad at\ t = 0 \tag{5.18}$$

5.2.1.4. Numerical Solution Program

The COMSOL Multiphysics is excellent, state-of-the-art software for the solution of many types of partial differential equations (PDEs), both stationary and time-dependent, by numerical techniques based on the finite element method for the spatial discretization. It also provides step-by-step instructions regarding how to solve a simple PDE using the COMSOL Multiphysics' graphical user interface (GUI). This example should be useful to get an initial idea regarding how COMSOL's GUI works, and it can also be used by users (or system administrators) to test whether a given installation of COMSOL works properly. Radio Frequency (RF) Module 3.4 is an optional package that extends the COMSOL Multiphysics® modeling environment with customized user interfaces and functionality optimized for the analysis of electromagnetic waves. Like all modules in the COMSOL Multiphysics family, it provides a library of prewritten ready-to-run models that make it quicker and easier to analyze discipline-specific problems. This particular module solves problems in the general field of electromagnetic waves, such as RF and microwave applications, optics, and photonics. The application modes (modeling interfaces) included here are fully multiphysics enabled, making it possible to couple them to any other physics application mode in COMSOL Multiphysics or the other modules.

The RF Module contains a set of application modes adapted to a broad category of electromagnetic simulations. Those who are not familiar with computational techniques but have a solid background in electromagnetics should find this module extremely beneficial. It can serve equally well as an excellent tool for educational purposes. Because the RF Module is smoothly integrated with all the COMSOL Multiphysics functionality, a simulation in this module can be coupled with an arbitrary simulation defined in any of the COMSOL Multiphysics application modes. This forms a powerful multiphysics model that solves all the equations simultaneously.

A guide to the COMSOL setting for this problem is shown in Figure 5.3. By using 3D Electro-Thermal Interaction mode, Figure 5.3, this mode predefines combinations of application modes and couplings for electro-thermal interaction.

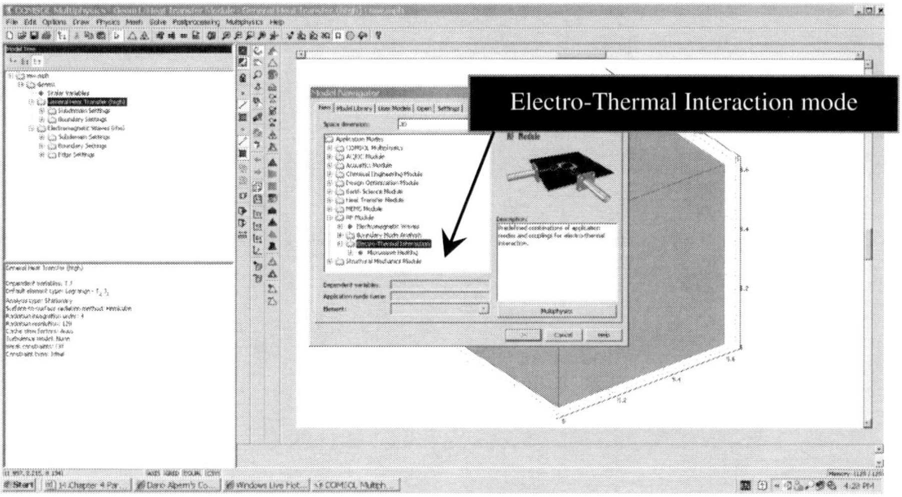

Figure 5.3. Mode predefines combinations of application modes and couplings for electro-thermal interaction.

5.2.1.5. Parameters Used to Calculate

The parameters used in this calculation are shown in the Table 5.2.

Table 5.2. Variables used for calculation

Variables	Values	Unit	Remark
Scalar			
Sigma_htgh (Stefan-Boltmann constant)	5.67e-0.08	$W.m^{-2}.deg^{-4}$	Data bases
Rg_htgh (Universal gas constant)	8.31451	$J.mol^{-1}.K^{-1}$	Data bases
Epsilon0_rfw (Permittivity of vacuum)	8.854187817e-12	$F.m^{-1}$	Data bases
Nu_rfwm (Frequency)	2.45e9	Hz	Data bases
General Heat Transfer (htgh)			
Subdomains: air, 1 atm			
k (isotropic)	0.0257	$W.m^{-1}.K^{-1}$	Data bases
ρ (density)	1.205	$kg. m^{-3}$	Data bases
C_p (heat capacity at constant pressure)	1.005	$J.kg^{-1}.K^{-1}$	Data bases
$T(t_0)$ Initial temperature	293.15	K	Test result
k (isotropic)	155	$W.m^{-1}.K^{-1}$	Data bases
ρ (density)	2730	$kg. m^{-3}$	Data bases

Table 5.2. (Continued)

Variables	Values	Unit	Remark
C_p (heat capacity at constant pressure)	893	$J.kg^{-1}.K^{-1}$	Data bases
$T(t_0)$ Initial temperature	293.15	K	Test result
Subdomains: Cement paste			
k (isotropic)	1.005-0.0025×T	$W.m^{-1}.K^{-1}$	Kim, 2003
ρ (density)	1915	$kg.\ m^{-3}$	Test result
C_p (heat capacity at constant pressure)	$0.75+3.43\times M_f^{water}$ $M_f^{water} = 0.38/1.38$	$J.kg^{-1}.K^{-1}$	Hansen, 1982
Q (heat source)	421.0559	$W.m^{-3}$	Test result
$T(t_0)$ Initial temperature	295.35	K	Test result
Boundary condition			
$T = T_0$	298.15	K	Test result
Electromagnetic Wave (rfw)			
Subdomains: air, 1 atm			
n (refractive index)	1.0008		Data bases
Subdomains: Aluminium 3003-H18			**Data bases**
relative permittivity	1		Data bases
electrical conductivity	2.326e7	$S.m^{-1}$	Data bases
Relative permeability	1		Data bases
Subdomains: Cement paste			**Data bases**
relative permittivity	15.0861-*j**4.4458		At 30 min delay time
electrical conductivity	1.042	$S.m^{-1}$	Test result
Relative permeability	1		Data bases
Subdomains: Aluminum 3003-H18			**Data bases**
Boundary condition			
Subdomains: air, 1 atm	Perfect electric conductivity		
Subdomains: Aluminum 3003-H18	Continuity		
Subdomains: Cement paste	Port, Power 390	watt	
Finite element			
Number of subdomains	7		
Dependent variables	Temperature (T)		
Mesh consists of elements.	84258	elements	
Analysis type of heat transfer	Transient		
Tolerance	0.01		
Transient duration	10, 20, and 30	min	
Sample	Cement paste		at w/c = 0.38

5.2.2. Results and Discussion

Based on the transient analysis (a microwave-cured sample's temperature-change rate is proportional to the difference between its own temperature and the temperature of its surroundings in the microwave cavity); the calculated temperature of the 1CW/C_P0.38 paste under microwave times of 10, 20 and 30 min. It illustrates an increase of maximum temperature within the samples as the microwave application times increase, such that temperatures of 61.27, 83.14, and 95.26°C were reached at times of 10, 20, and 30 min, respectively. In addition, the points at which maximum and minimum temperatures occur are in the middle of the sample and at the edge of an inlet waveguide. The maximum temperature point arises from the interaction between microwave energy and cement paste; the heat generated by the microwave energy works in conjunction with the heat generated by the hydration of the cement, so that at the center the accumulation is high and very little heat dissipates.

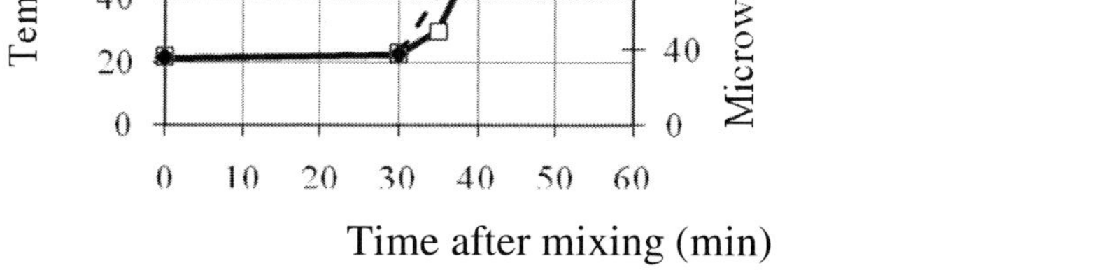

Figure 5.4. Temperature history of the 1CW/C_0.38 paste when applying microwave power 100 watt on a period of 30 min.

The predictions and experimental data of the maximum temperature history within the 1CW/C_0.38 paste when microwave power of 100 watt is applied for up to 30 min are compared in Figure 5.4. The result shows that the prediction data are higher than those of the experimental data. This is mainly due to the fact that the parameter used in this calculation, especially the permittivity value, is kept constant at $\varepsilon_r^* = 15.0861 - \vec{j}4.4458$, while in the real state this value decreases continuously as free water content decreases. That is, free water content is consumed and then transformed to fixed water (Taylor, 1997); as a result, microwave radiation can interact with cement paste and generate a low amount of heat. As well as the dielectric value, the heat transfer values also changes according to how long microwave energy is applied and the formation of hydration products decreases heat transfer with time also. Furthermore, it can be seen that heat generation takes place when the free water can move from the center to another location within the paste; therefore, at the middle point, the temperature is then dropped.

5.3. Conclusion

In order to predict a temperature rise during curing by microwave energy, the model explored the interaction between the microwave (electromagnetic field) and cement paste, the heat dissipation (conduction mode) model by using the Multiphysics Modeling and Simulation: COMSOL. The results showed that the simple model associated with assumptions, and initial and boundary conditions can predict a temperature rise higher than those of the experimental data.

Chapter 6

CONCLUSIONS AND RECOMMENDATIONS FOR FUTURE STUDY

6.1. CONCLUSIONS

The objectives of this research were to (a) investigate the mechanisms of changes in the dielectric properties of plain containing pozzolan materials and reinforced concrete, (b) characterize the microstructure of microwave-cured plain and reinforced concrete, (c) study the mechanism of microwave heat generation, heat dissipation, and mass transfer of plain containing pozzolan materials and reinforced concrete, and (d) formulate mathematical models for the relationships between microwave curing and mass transfer for steel-reinforced concrete.

In order to achieve the stated goals, detailed experimental and computational methodologies were used. The method experimentally studied the interaction between microwave energy and Type I reinforced cement pastes during the first 24-h hydration period. The second method developed a theoretical microwave-heating model to achieve high-temperature curing using microwave energy in associated a single-mode rectangular waveguide.

The microwave system used to transfer microwave energy in order to heat or cure the reinforced pastes, mortars and concretes consisted of the following:

- The microwave system used was a monochromatic microwave at a frequency of 2.45 GHz.
- Microwave energy was generated by a magnetron and transmitted directly along the propagation direction (+z) of a rectangular wave guide toward a water load situated at the end of the waveguide to ensure that a minimal amount of microwave energy would be reflected back to the sample.
- A warming water load was circulated through the cooling tower in order to reduce the temperature in the water load system.
- A sample was arranged perpendicular to the propagation direction.
- A Type K thermo-couple with a 0.1 mm diameter was inserted at the center of the sample for the purpose of monitoring the temperature rise.
- During a 10-min period of microwave heating, the output of the microwave magnetron was controlled at 1000 W. The microwave plane wave traveled directly along the wave guide and made contact with the sample surface; the wave was then reflected and transmitted. By using a wattmeter, incident, reflected and transmitted waves were monitored.

It can be summarized that:

6.1.1. Dielectric Permittivity of Reinforced Cement-Based Materials

The dielectric permittivity of reinforced cement-based materials is affected by the initial water-cement mass ratio (w/c), the cement type, the presence/lack of pozzolan, and the aggregate type. It should be noted here that though the volumetric fraction of water and the superplasticizer of a given mixture may be small, they strongly affect the dielectric permittivity of the cement. This is because both water and the superplasticizer have high dielectric permittivity. The change in the dielectric permittivity of cement-based materials, therefore, is relatively high and remains constant during the dormant period. However, the change in the dielectric permittivity decreases rapidly when the hydration reaction resumes, and it continues to decrease during the acceleratory period.

6.1.2. Microwave-Assisted Curing of Pastes and Concretes

The study took into account microwave energy's ability to accelerate hydration reactions and increases in temperature within the microwave-cured reforced sample, as well as the effects of water-to-solid mass ratios, pozzolan materials, mortar and concrete, delay times, power level, application duration, sequential process, and the environment's normal and low pressure. Furthermore, the characteristics of microwave-cured specimens and normal (lime-saturated deionized water) and autoclave curing were compared in detail. The test results showed that the temperature increased monotonically among the positions of measurement during the microwave-curing process. The typical micrographs of the microwave-cured paste at the age of 4 h after mixing, 28 days after curing in lime-saturated deionized water, and when subjected to microwave energy showed that the samples consisted of hydrated phases and pores, as well as cores of $Ca(OH)_2$ dendrite crystals or other crystals (marked CH), C–S–H, and granular structure. Furthermore, some ettringite (Aft) was found in the samples cured by microwave energy. From the SEM–EDS results, it was observed that although the measured Ca/Si ratios of the pastes were similar in magnitude, they consistently decreased when the temperature decreased. For compressive strengths, the microwave-cured pastes developed strength quite rapidly in accord with the maintenance of the level of microwave power and time of application. In addition, the water content of the pastes' hydration products was affected depending on the w/c, and the $Ca(OH)_2$ contents were affected due to the completion of the hydration reaction.

6.1.3. A Theoretical Model to Predict the Heat Transfer of Reinforced Cement Paste Subjected to Microwave Energy

To predict the extent to which temperature will increase, the model explored the interactions between the microwave (electromagnetic field) and cement paste, that is, the heat dissipation (conduction mode) model, by using Multiphysics Modeling and Simulation: COMSOL. The modeling results showed that the simple model, that is, a specific physical model, with assumptions and initial and boundary conditions, can predict some aspects of temperature increases.

6.2. Recommendations for Future Study

Given that current trends in clean energy are emphasizing sustainable development and environmentally friendly processes, microwave energy is gradually becoming a more popular energy source for heating and drying in various industrial applications. However, microwave energy has been put to such use only very gradually because of a lack of basic knowledge and serious concerns regarding radiation safety and hazards to biological tissue. Thus, these issues together with the socio-technical questioning of this method need to be clarified.

Still, in the future, there will be industrial-scale applications for which conventional curing methods will not be appropriate. Therefore, further issues are the compatibility of mixed concretes and curing conditions, heat- and mass-transfer mechanisms, and numerical tools to predict heat and mass transfer during heating by microwave energy. The use of microwave energy to improve the properties of cement-based materials is a relatively new area of research that is attracting significant interest because microwave heating has many advantages, among which are rapid heat generation, high-energy penetration, instantaneous and precise electronic control, and clean process. In addition, conventional curing methods have many limitations. For example, they take a long time to reach the strength required for water curing. In addition, due to the inherent thermal insulation of concrete, their hydration products are non-uniform, which causes different temperatures to occur in processed concrete under high-stream and temperature-curing conditions. Some essential points that illustrate the performance of microwave energy are as follows:

The second point concerns the heat- and mass-transfer mechanisms within the heated concrete during microwave curing. A number of studies have shown that microwave curing improves the generation and transportation of heat and mass within the heated concrete material, which are key performance factors in regard to energy efficiency. This is due to microwave energy's inherent ability to volumetrically generate heat from the inside to the outside. However, the assumption that no chemical reactions occur within the concrete is incorrect. The hydration and pozzolanic reactions take place naturally; therefore, the difficulty may be a definitive point in the development of microwave-curing in determining when and how to apply microwave energy in a way that is appropriate for achieving a particular reaction in the cement.

The third point concerns the numerical tools used to predict heat and mass transfer during heating using microwave energy. This study developed a model to account for a range of aspects. Further, the study evaluated variations in temperature as they related to the heating time, frequency, and sample size of the cement paste with water-to-cement ratios (w/c) with weights of 0.25, 0.38, and 0.45. The model developed here is an example of efficient computational prototyping.

The last point concerns the compatibility of microwave-heating mechanisms and the suitability of the equipment and systems associated with them. In other words, these components should be closely compatible with the concrete material to be processed. Heating mechanisms that take place inside a multi-mode cavity are somewhat more complicated than mechanisms that occur in a single-mode cavity. However, multi-mode microwave systems are generally used in industry. The way to solve this problem is to extrapolate the results of a single mode and then predict the phenomena that occur in a multi-mode cavity as accurately as possible.

REFERENCES

Abdelghani-Idrissi MA, Experimental investigations of occupied volume effect on the microwave heating and drying kinetics of cement powder in a mono-mode cavity. Appl Therm Eng 2001; 21: 955-65.

Agrawal DK, Latest global developments in microwave materials processing. Mat Res Inno 2011; 14: 3-8.

Al-Qadi IL, Hazim O, Su W, Riad S. Dielectric properties of Portland cement concrete at low radio frequencies. J Mater Civ Eng 1995; 7: 192-8.

Ayappa KG, Davis HT, Crapiste G, Davis EAJ. Gordon,Microwave Heating: An Evaluation of Power Formulations. Chem Engng Sci 1991; 46: 1005-16.

Ayappa KG, Davis HT, Davis EA, Gordon J. Analysis of Microwave Heating of Materials with Temperature-Dependent Properties. AIChE J 1991; 37: 313-22.

Baoyi L, Yuping D, Shunhua L. The electromagnetic characteristics of fly ash and absorbing properties of cement-based composites using fly ash as cement replacement. Constr Build Mater 2012; 27: 184-8.

Bažant ZP, Zi G. Decontamination of Radionuclides from Concrete by Microwave Heating. I: Theory. J Eng Mech 2003; 129: 777-84.

Bažant, Z.P., Zi, G. (2003). Decontamination of Radionuclides from Concrete by Microwave Heating I: Theory. Journal of Engineering Mechanics, 129(6): pp. 777–784.

Bendsted, J., Barnes, P. (2002). Structure and Performance of Cement. Spon Press: London.

Bois KA, Benally A, Zoughi R, Near-field microwave non-invasive determination of NaCl in mortar. IEEE proceedings - Science, Measurement and Technology, Special Issue on Non-destructive Testing and Evaluation 2001; 148: 178-82.

Bois KA, Benally A, Zoughi R. Microwave near-field reflection property analysis of concrete for material content determination. IEEE Trans Instrum Meas 2000; 49: 49-55.

Büyüköztürk O, Yu T, Ortega J. A methodology for determining complex permittivity of construction materials based on transmission-only coherent, wide-bandwidth free-space measurements. Cem Conc Res 2006; 8: 349-59.

Coverdale RT, Christensen BJ, Mason TO, Jennings HM, Garboczi EJ. Interpretation of impedance spectroscopy of cement paste via computer modeling: Part II. J Mater Sci 1994; 29: 4984-92.

Cronin NJ. Microwave and Optical Waveguides. London; Taylor & Francis; 1995.

Davis, R.E., Carlson, R.W., Kelly, J.W., Davis, A.G. (1937). Properties of Cements and Concretes Containing Fly Ash. Proc. American Concrete Institute 33: pp. 577–612

Dikhtiar V, Jerby E. Patent No. US 6114676, Method and device for drilling, cutting, nailing and joining solid non-conductive materials using microwave radiation, issued in 2000.

Ding XZ, Zhang X, Ong CK, Tan BT. Study of dielectric and electrical properties of mortar in the early hydration period at microwave frequencies. J Mater Sci 1996; 31: 5339-45.

Dongxu L, Xuequan W. A study on the application of vacuum microwave composite dewatering technnique in concrete engineering. Cem Concr Res 1994; 24: 159-64.

Donnell KM, Hatfield S, Zoughi R, Kurtis KE. Wideband microwave characterization of alkali-silica reaction (ASR) gel in cement-based materials. Mat Lett 2013; 90: 159-61.

Donnell KM, Zoughi R, Kurtis KE. Demonstration of microwave method for detection of alkali–silica reaction (ASR) gel in cement-based materials. Cem Conc Res 2013; 44: 1-7.

Ebadian MA, Li W, A theoretical/experimental investigation of the decontamination of a radioactively contaminated concrete surface using microwave technology: Final Report. DOE Project, DE-AC05-840R2140; 1992.

Energy efficiency asia, http://www.energyefficiencyasia.org/docs/Industry Sectors Cement draft_May 05.pdf, 20/08/2010; 2010.

Engelbrecht HCL, Birch-Rasmussen S. Patent No. EP2197641B1, Process for curing and drying reinforced concrete, Publication date: Feb 23, 2011.

Engelbrecht HCL, Birch-Rasmussen S. Patent No. WO2009027813A2, Process for curing and drying reinforced concrete, Publication date: Mar 5, 2009.

Fang Y, Chen Y, Silsbee MR, Roy DM, Microwave sintering of fly ash. Mater Lett 1996; 27: 155-59.

Fang Y, Roy DM, Roy R. Microwave clinkering of ordinary and colored Portland cements. Cem Conc Res 1996; 26: 41-7.

Feng, L., Meyer, C. (2009). Hydration Kinetics Modeling of Portland Cement Considering the Effects of Curing Temperature and Applied Pressure. Cement and Concrete Research, 39: pp. 255–265.

Gorur K, Smit MK, Wittmann FH. Microwave study of hydrating cement paste at Early Age. Cem Concr Res 1982; 12: 447-54.

Haddad RH, Ai-Qadi IL. Characterization of Portland cement concrete using electromagnetic waves over the microwave frequencies. Cem Conc Res 1998; 28: 1379-91.

Hager NE, Domszy RC. Monitoring of cement hydration by broadband time-domain-reflectometry dielectric spectroscopy. J Appl Phys 2004; 96: 5117-28.

Haoxuan L, Agrawal DK, Cheng J, Silsbee MR. Formation and hydration of C3S prepared by microwave and conventional sintering. Cem Conc Res 1996; 29: 1611-17.

Haoxuan L, Agrawal DK, Cheng J, Silsbee MR. Microwave sintering of sulphoaluminate cement with utility wastes. Cem Conc Res 2001; 31: 1257-61.

Hashem M, Al-Mattarneh A, Ghodgaonkar DK, Mahmood W, Majid WA. Determination of compressive strength of concrete using free-space reflection measurements in the frequency range of 8-12.5 GHz. Asia-Pacific Microwave Conference, Taipei, Taiwan; 2001: 679-82.

Hasted JB, Shah MA, Microwave absorption by water in building materials. British J Appl Phys 1964; 15: 825, doi:10.1088/0508-3443/15/7/307.

Henry F, Broncy M, Berteaud AJ.The hydration kinetics studied by means of microwaves. Microwaves 1978; 19: 608-12.

Hewlett Packard Corporation. Dielectric Probe Kit 85070A. Palo Alto. CA: Research and Development Unit. Test and Measurements Laboratories; 1992.

Hills DL, The removal of concrete layers from biological shields by microwave EUR 12185 ((2nd Edn).). Nuclear Science and Technology, Commission of the European Communities; 1989.

Hippel ARV. Dielectric Materials and Applications. New York: The Technology Press of M.I.T. and John Wiley & Sons; 1954.

Hutchinson RG, Chang JT, Jennings HM, Brodwin ME. Thermal acceleration of Portland cement mortars with microwave energy. Cem Concr Res 1991; 21: 795-9.

Institute of Electrical and Electronics Engineers (IEEE). Standard for Safety Levels with Respect to Human Exposure to Radio Frequency Electromagnetic Fields 3 kHz to 300 GHz. IEEE/ANSI, C95.1; 2005.

Institute of Electrical and Electronics Engineers (IEEE). Technical information statement on: Human Exposure to Microwaves and Other Radio Frequency Electromagnetic Fields. IEEE Eng Med Biol Mag 1995; 14: 336-7.

International Radiation Protection Association (IRPA). International commission on non-ionizing radiation protection (ICNIRP) Guidelines: For limiting exposure to time-varying electric, magnetic and electromagnetic fields (up to 300 GHz), Health Physics 1988; 54: 492-522.

Jerby E, Dikhtyar V, Aktushev O, GrosglickU. The microwave drill. Science 2002; 298 : 587-9.

Jiang H, Hao Q, Zhou J, Microwave synthesis of dicalcium silicate: -Comparison with conventional synthesis. Adv Mat Res 2011; 306-307: 1060-67.

Ke K, Ma B, Wang X. Formation of tricalcium silicate prepared by electric and microwave sintering. Adv Mat Res 2011; 148-149: 1119-23.

Kjellsen O. Knut, Detwiler J. Rachel, Gjørv, E. Odd. (1990). Pore Structure of Plain Cement Pastes Hydrated at Different Temperatures. Cement and Concrete Research, 20: pp. 927–933.

Klysz G, Balayssac JP, Ferrières X. Evaluation of dielectric properties of concrete by a numerical FDTD model of a GPR coupled antenna-Parametric study. NDT & E Int 2008; 41: 621-31.

Klysz G, Balayssac JP. Determination of volumetric water content of concrete using ground-penetrating radar. Cem Conc Res 2007; 37: 1164-71.

Lagos LE, Li W, Ebadian MA, White TL, Grubb RG, Foster D. Heat transfer within a concrete slab with a finite microwave heating source. Int J Heat Mass Transfer 1995; 38: 887-97.

Lee M. Preliminary study for strength and freeze-thaw durability of microwave- and steam-cured concrete. J Mater Civ Eng 2007; 19: 972-6.

Lee M. Preliminary study for strength and freeze-thaw durability of microwave- and steam-cured concrete. J Mater Civ Eng 2007; 19: 972-6.

Leon CO. Effect of Mixture Composition and Time on Dielectric Constant of Fresh Concrete. Master Thesis. North Carolina State University: 2007.

Leung CKY, Pheeraphan T. Determination of optimal process for microwave curing of concrete. Cem Concr Res 1997; 27: 463-72.

Leung CKY, Pheeraphan T. Microwave curing of Portland cement concrete: experimental results and feasibility for practical applications. Constr Build Mater 1995; 9: 67-73.

Leung CKY, Pheeraphan T. Very high early strength of microwave cured concrete. Cem Concr Res 1995; 25: 136-46.

Levita G, Marchetti A, Gallone G, Princigallo A, Guerrini GL. Electrical properties of fluidified Portland cement mixes in the early stage of hydration. Cem Conc Res 2000; 30: 923-30.

Li W, Ebadian MA, White TL, Grubb RG, Foster D, Heat and mass transfer in a contaminated porous concrete slab subjected to microwave heating General Papers in Heat Transfer and Heat Transfer in Hazardous Waste. Processing, ASME HTD ((2nd Edn).) 1992; 212: 143-53.

Li W, Ebadian MA, White TL, Grubb RG, Foster D. Heat and mass transfer in a contaminated porous concrete slab with variable dielectric properties. Int J Heat Mass Transfe 1994; 37: 1013-27.

Li W, Ebadian MA, White TL, Grubb RG, Foster D. Heat transfer within a radioactive contaminated concrete slab applying a microwave heating technique. ASME Trans J Heat Transfer 1993; 115: 42-50.

Li W, White TL, Foster D, Ebadian MA. Heat transfer within a steel-reinforced porous concrete slab subjected to microwave heating. J Heat Transfer 1995; 117(3): 582-9.

Ludewig A., Steinbach D. Patent No. WO1997021060A1, Method and device for drying out buildings and or fixed components, Publication date: Jun 12, 1997.

Madlool NA, Saidur R, Hossaina MS, Rahim NA. A critical review on energy use and savings in the cement industries. Renew Sust Energ Rev 2011; 15: 2042-2060.

Mak SL, Banks R, Ritchie R, Shapiro G. Practical industrial microwave technology for rapid curing of precast concrete. Presented to Concrete 2001: 20th Biennial Concrete Conf., Perth, Western Australia, 11-14 September: 2001.

Mak SL, Banks RW, Ritchie DJ, Shapiro G. Advances in microwave curing of concrete. Presented to 4th World Congress on Microwave & Radio Frequency Applications, Sydney, Australia, 22-26 September: 2002.

Mak SL, Ritchie DJ, Shapiro G, Banks RW, Rapid microwave curing of precast concrete slab elements. Presented to 5th CANMET/ACI Int. Conf. on Recent Advances in Concrete Technology, Singapore, 29 July to 1 August: 2001.

Mak SL, Shapiro TGS. Accelerated heating of concrete with microwave curing. Proc. 4th CANMET/ACI/JCI Int. Conf. on Recent Advances in Concrete Technology: Tokushima, Japan, 7-11 June 1989: 531-42.

Mak SL, Taylor AH, Son T, El-Hassan MT. Performance of concrete subjected to microwave accelerated processing. Proc. 4th CANMET/ACI Int. Conf. on Durability of Concrete, Sydney, Australia, 17-22 August 1997: 603-15.

Mak SL. Microwave accelerated processing for precast concrete production. Proc. 4th CANMET/ACI Int. Conf. on Durability of Concrete, Sydney, Australia, 17-22 Aug., supplementary papers 1997: 709-20.

Mak SL. Properties of heat cured concrete. Presented to Concrete Institute of Australia Seminar on Heat, Fire, Weather Effects on Concrete. Brisbane; Queensland; Australia: April, 1999: 156-63.

Makul N, Agrawal DK, Influences of microwave-accelerated curing procedures on microstructure and strength characterization of Type I-Portland cement pastes. J Cer Pro Res 2012; 13: 376-81.

Makul N, Agrawal DK. Comparison of the microstructure and compressive strength of Type 1 Portland cement paste between accelerated curing methods by microwave energy and autoclaving, and a saturated-lime deionized water curing method. J Cer Pro Res 2012; 13: 174-7.

Makul N, Agrawal DK. Influence of microwave-accelerated curing procedures on the microstructure and strength characteristics of type I-Portland cement pastes. J Cer Pro Res 2011; 12: 376-81.

Makul N, Agrawal DK. Microwave (2.45 GHz)-assisted rapid sintering of SiO2-rich rice husk ash. Mat Lett 2010; 64: 367-70.

Makul N, Agrawal DK. Microwave-accelerated curing of cement-based materials: Compressive strength and maturity modeling. Key Eng Mater 2011; 484: 210-221.

Makul N, Keangin P, Rattanadecho P, Chatveera B, Agrawal DK. Microwave-assisted heating of cementitious materials: Relative dielectric properties, mechanical property, and experimental and numerical heat transfer characteristics. Int Commun Heat Mass Transfer 2010; 37: 1096-105.

Makul N, Rattanadecho P, Agrawal DK. Microwave curing at an operating frequency of 2.45 GHz of Portland cement paste at early-stage using a multi-mode cavity: Experimental and numerical analysis on heat transfer characteristics. Int Commun Heat Mass Transfer 2010; 37: 1487-1495.

Mario P., Sergio L. Patent No. EP0462612A1, Process and device for accelerating the drying of cement mixes, Publication date: Dec 27, 1991.

Meducin, F., Zanni, H., Noik, C., Hamel, G., Bresson, B. (2008). Tricalcium Silicate (C3S) Hydration under High Pressure at Ambient and High Temperature (200 °C). Cement and Concrete Research, 38(3): pp. 320–324.

Metaxas AC, Meredith RJ. Industrial Microwave Heating. United Kingdom: Peter Peregrinus; Herts; 1998.

Metaxas AC. Microwave Heating. J Microwave Power Electromagn Energy 1991; 5: 237-47.

Mindress, S., Young, J.F., Darwin, D. (2002). Concrete. Prentice Hall: Sidney, Australia.

Moukwa M, Brodwin M, Christo S, Chang J, Shah SP. The influence of the hydration process upon microwave properties of cements. Cem Concr Res 1991; 21: 863-72.

Nagi M, Whiting D. Determination of water content of fresh concrete using a microwave oven. Cem Conc Aggre 1994; 16: 125-31.

Naik TR, Ramme BW. Determination of the water content of concrete by the microwave method. Cem Concr Res 1987; 17: 927-38.

National Radiological Protection Board (NRPB) and the Health Protection Agency. Review of the Scientific Evidence for Limiting Exposure to Electromagnetic Fields (0-300 GHz). 2004.

Neelakantan TR, Ramasundaram S, Shanmugavel R. Prediction of 28-day compressive strength of concrete from early strength and accelerated curing parameters. Int J Eng Tech 2013; 5: 1197-201.

Neville AM, Properties of Concrete, Fourth Edition, London England: Pitman Books Limited; 1995.

Occupational Health and Safety Administration (OSHA). Radiofrequency and microwave radiation, 2009. www.osha.gov/SLTC/radiofrequency radiation/ (Access date: Dec 4th, 2013)

Odler, I., Abdul-Maula, S. (1987). Investigations on the Relationship between Porosity Structure and Strength of Hydrated Portland Cement Pastes III: Effect of Clinker Composition and Gypsum Addition. Cement and Concrete Research, 17: pp. 22–30.

Oriol M, Pera J. Pozzolanic activity of metakaolin under microwave treatment. Cem Concr Res 1995; 25: 265-70.

Paul C, Stephen B. Dielectric properties of Portland cement paste as a function of time since mixing. J Appl Phys 1989; 66: 6007-13.

Pheeraphan T, Accelerated Curing of Concrete with Microwave Energy. Doctor of Philosophy Dissertation; MIT; 1997.

Pheeraphan T, Leung CKY. Freeze-thaw durability of microwave cured air-entrained concrete. Cem Concr Res 1997; 27: 427-35.

Pheeraphan T, Leung CKY. Freeze-thaw durability of microwave cured air-entrained concrete. Cem Concr Res 1997; 27: 427-35.

Pheeraphana T, Cayliani L, Dumangas Jr MI, Nimityongskul P. Prediction of later-age compressive strength of normal concrete based on the accelerated strength of concrete cured with microwave energy. Cem Conc Res 2002; 32: 521-7.

Ping G, Beudoin JJ. Dielectric behaviour of haredened cement paste systems. J Mater Sci Lett 1996; 15: 182-4.

Prasad A, Prasad K. Effective permittivity of random composite media: a comparative study. Physica B 2007; 396: 132-7.

Quéméneur L, Choisnet J, Raveau B, Thiebaut JM, Roussy G. Microwave clinkering with a grooved resonant applicator. J Am Ceram Soc 1983; 66: 855-9.

Quéméneur L, Choisnet J, Raveau B. Is it possible to use the microwave for clinkering the cement raw materials?. Mater Chem Phys 1983; 8: 293-303.

Ramezanianpour AA, Khazali MH, Vosoughi P. Effect of steam curing cycles on strength and durability of SCC: A case study in precast concrete. Constr Build Mater 2013; 49: 807-13.

Rattanadecho P, Suwannapum N, Chatveera B, Atong D, Makul N. Development of compressive strength of cement paste under accelerated curing by using a continuous microwave thermal processor. Mater Sci Eng A 2008; 472: 299-307.

Rattanadecho P, Aoki K, Akahori M. A Numerical and Experimental Investigation of the Modeling of Microwave Drying Using a Rectangular Wave Guide. Drying Technol An Int J 2001; 19: 2209-34.

Rattanadecho P, Suwannapum N, Cha-um W. Interactions Between Electromagnetic and Thermal Fields in Microwave Heating of Hardened Type -Cement Paste Using a Rectangular Waveguide (Influence of Frequency and Sample Size). ASME J Heat Transfer 2009; 131: 1-12.

Reboul JP. The hydraulic reaction of tricalcium silicate observed by microwave dielectric measurements. Rev Phys Appl (Paris); 1978; 13: 383-6.

Rhim HC, Büyüköztürk O. Electromagnetic properties of concrete at microwave frequency

Rhim HC, Büyüköztürk O. Electromagnetic properties of concrete at microwave frequency range. ACI Mater J 1998; 95-M25: 262-71.

Richardson, I.G., Brydson, R., McComb, D.W., Groves, G.W. (1993). Parallel Electron Energy Loss Spectroscopy Study of Al-substituted Calcium Silicate Hydrate (C–S–H) Phases Present in Hardened Cement Pastes. Solid State Communications, 88(2): pp. 183–187.

Sohn D, Johnson DL. Microwave curing effects on the 28-day strength of cementitious materials. Cem Concr Res 1999; 29: 241-7.

Somaratna J, Ravikumar D, Neithalath N. Response of alkali activated fly ash mortars to microwave curing. Cem Concr Res 2010; 40: 1688-96.

Soustos MN, Bungey JH, Millard SG, Shaw MR, Patterson A. Dielectric properties of concrete and their influence on radar testing. NDT&E Int 2001; 34: 419-25.

Suston WH, Microwave processing of ceramic materials. Am Ceram Soc Soc 1989; 68: 3-19.

Suwannapum N, Rattanadecho P. Analysis of heat-mass transport and pressure buildup induced inside unsaturated porous media subjected to microwave energy using a single (TE10) mode cavity. Drying Technol Int J 2011; 29: 1010-24.

Taylor, H.F.W. (1986). Proposed Structure for Calcium Silicate Hydrate Gel. Journal of the American Ceramic Society, 69: pp. 464–467.

Taylor, H.F.W. (1997). Cement Chemistry; Thomas Telford Publishing: London, Great Britain.

Tereshchenko OV, Buesink FJK, Leferink FBJ. An overview of the techniques for measuring the dielectric properties of materials. 978-1-4244-6051-9/11/ IEEE; 2011.

Venkatesh MS, Raghevan GSV. An overview of dielectric properties measuring techniques. Can Biosyst Eng 2005; 47: 15-29.

Verbeck GJ, Helmuth RA. Structures and physical properties of cement paste. 5th Int. Congress Cement Chemistry: Tokyo; Japan; 1969: 1-44.

Vongpradubchai S, Rattanadecho P. The Microwave processing of wood using a continuous microwave belt drier. Chem Eng Process Process Intensif 2009; 48: 997-1003.

Watson A. Building research station Report A 93. England (DSIR): Garston; 1961.

Watson A. Curing of Concrete. E.C. Okress (Ed.). Microwave Power Engineering. New York: Academic Press; 1968.

Watson A. The non-destructive measurement of water content by microwave absorption. C.I.B. No. 3; 1960: 15-6.

Wen S, Chung DDL. Effect of admixtures on the dielectric constant of cement Paste. Cem Conc Res 2001; 31: 673-7.

White TL, Grubb RG, Pugh LP, Foster D Jr, Box WD. Removal of contaminated concrete surfaces by microwave heating—phase I results presented at the 18th American Nuclear Society Symposium on Waste Management Waste Management 92, Tucson, Arizona; 1992.

Wittmann FH, Schlude F. Microwave absorption of hardened cement paste. Cem Concr Res 1975; 5: 63-71.

Xuequan W, Jianbgo D, Mingshu T, Microwave curing technique in concrete manufacture. Cem Conc Res 1987; 17: 205-10.

Yasunaka, H., Shibamoto, M., Sukagawa, T. 1987. Microwave decontaminator for concrete surface decontamination in JPDR. Proceedings of the International Decommissioning Symposium. United States of America. 4-109- 116.

Youssef H, Smith A, Bonnet JP, Abelard P, Blanchart P. Electrical characterization of aluminous cement at the early age in the 10 Hz - 1 GHz frequency range. Cem Conc Res 2000; 30: 1057-62.

Zhang J, Scherer GW. Comparison of methods for arresting hydration of cement. Cem Conc Res 2011; 41: 1024-36.

Zhang X, Ding XZ, Lim TH, Ong CK, Tan BTG, Yang J. Microwave study of hydration of slag cement blends in early period. Cem Conc Res 1995; 25: 1086-94.

Zi G, Bažant ZP. Decontamination of Radionuclides from Concrete by Microwave Heating. II: Computations. J Eng Mech: 2003; 129: 785-92.

Zoughi R, Gray SD, Nowak PS. Microwave nondestructive estimation of cement paste compressive strength. ACI Mat J 1995; 92: 64-70.

Zoughi R, Microwave Non-Destructive Testing and Evaluation. Chapter 2, The Netherlands: Kluwer Academic Publishers; 2000.

Zoughi R, Nowak PS, Bois KJ, Benally AD, Mirshahi R, Campbell H, Near-field microwave inspection of cement based materials - microwave sensor for nondestructive and non-contact estimation of concrete compressive strength, Final Report. NSF Contract no. CMS-9523264 and EPRI Contact no. WO 8031-09; 1998.

INDEX

A

B

C

D

E

F

H

I

L

M

V

W

X